AutoCAD® Pock

Reference
2007 Edition

Covers Releases

2007, 2006, 2005 and 2004

by
Cheryl R. Shrock

Professor
Drafting Technology
Orange Coast College, Costa Mesa, Ca.
and
Autodesk® Authorized Author and Autodesk® Authorized Publisher

INDUSTRIAL PRESS
New York

Industrial Press Inc.
989 Avenue of the Americas
New York, NY 10018

10 9 8 7 6 5 4 3 2 1

Why do you need this book?

Refresh your memory or learn something new.
No need to memorize.
Handy size and easy to use.

About this book

The *AutoCAD® Pocket Reference* includes all the important fundamental
Commands, Concepts, and How to information for the every day use of
AutoCAD. It is not designed to take the place of larger textbooks but rather
to supplement them as a quick reference.

How to use this book

The information in this book has been organized in 12 sections.
Each section contains related material. For example, if you needed
information regarding dimensioning, you would go to:
Section 3: Dimensioning.
A comprehensive Table of Contents and a cross-referenced Index have
been carefully prepared to insure easy access to all information.

Words from the Author

*"I originally wrote this book for myself. Occasionally I forget commands and
"How to" steps also. It's convenient small size allows me to toss it in my
briefcase or set it beside my monitor. No need to memorize anymore."*

Cheryl R. Shrock is a Professor and Chairperson of Computer Aided
Design at Orange Coast College in Costa Mesa, California.
She is also an Autodesk® authorized
author / publisher and has written 14 versions of
"Exercise Workbook for Beginning and Advanced AutoCAD".
Cheryl is always trying to think of new ways to make it easy to learn
AutoCAD. This Pocket Reference is the latest in that endeavor.

Exercise Workbooks written by Cheryl R. Shrock and available from Industrial Press:

Beginning AutoCAD **2004**

ISBN 0-8311-3198-5 or
978-08311-31982

Advanced AutoCAD **2004**

ISBN 0-8311-3199-3 or
978-08311-31999

Beginning AutoCAD **2005**

ISBN 0-8311-3200-0 or
978-08311-32002

Advanced AutoCAD **2005**

ISBN 0-8311-3201-9 or
978-08311-32019

Beginning AutoCAD **2006**

ISBN 0-8311-3213-2 or
978-08311-32132

Advanced AutoCAD **2006**

ISBN 0-8311-3214-0 or
978-08311-32149

Beginning AutoCAD **2007**

ISBN 0-8311-3302-3 or
978-08311-33023

Advanced AutoCAD **2007**

ISBN 0-8311-3303-1 or
978-08311-33030

*For information about these workbooks,
visit www.industrialpress.com*

*For information about Cheryl Shrock's online courses,
visit www.shrockpublishing.com*

TABLE OF CONTENTS

The following Table of Contents is an overview of the contents within this book.
Refer to the Index at the back of this book to easily locate specific information.

SECTION 1

Action Commands

Array	1-2
Break	1-6
Chamfer	1-8
Copy	1-10
Divide	1-11
Erase	1-12
Explode	1-13
Extend	1-14
Fillet	1-15
Inquiry	1-17
Match Properties	1-18
Measure	1-20
Move	1-21
Offset	1-22
Rotate	1-24
Scale	1-25
Stretch	1-26
Trim	1-27
Undo/Redo	1-28
Wipeout	1-29
Zoom	1-30

SECTION 2

Concepts

Model / Paper space	2-2
Creating scaled dwgs	2-3

SECTION 3

Dimensioning

Dimensioning	3-2
Ordinate dim's	3-17
Tolerances	3-21
Geometric Tolerancing	3-23
Modify Dimensions	3-28
Dimension Styles	3-35
Definitions	3-44
Trans-spatial	3-49

SECTION 4

Drawing Entities

Arc	4-2
Blocks	4-6
Centermark	4-25
Circle	4-26
Donut	4-28
Ellipse	4-29
Gradient Fills	4-30
Hatch	4-32
Lines	4-42
Point	4-43
Polygon	4-44
Polylines	4-45
Rectangle	4-48
Revision Cloud	4-50

SECTION 5

How to...

Add a Printer	5-2
Create a Field	5-41
Edit a Field	5-43
Create a Page Setup-2004	5-6
Create a Page Setup 06/05	5-9
Create a Table	5-33
Insert a Table	5-35
Modify a Table	5-39
Create a Template	5-13
Open a template	5-15
Open an existing file	5-16
Exit AutoCAD	5-17
Create a Viewport	5-18
Customize Mouse	5-22
Calculate DSF	5-23
Lineweights to colors	5-27
Save a Drawing	5-30
Select objects	5-31
Start a New Drawing	5-32

SECTION 6

Layers

Selecting Layers	6-2
Creating new (2004)	6-5
Creating new (06/05)	6-6
Load Linetypes (04)	6-8
Load Linetypes (06/05)	6-9
Layer control def's (04)	6-10
Layer control def's (06/05)	6-11

SECTION 7

Input options

Coordinate Input	7-2
Direct Distance Entry	7-5
Dynamic Input	7-6
Polar Coordinates	7-11
Polar Snap	7-12
Polar Tracking	7-13

SECTION 8

Miscellaneous

Background Mask	8-2
Backup Files	8-3
Grips	8-4
Object Snap	8-5
Pan	8-9
Properties Palette	8-11
Metric Conversion Factors	8-12
Drawing Scales	8-13
Print Quick Letter size Draft	8-15

SECTION 9

Plotting

Background Plotting	9-2
Plotting, Model tab (2004)	9-3
Plotting, Model tab (06/05)	9-5
Plotting, Layout tab (04)	9-9
Plotting, Layout tab (6 & 5)	9-13

SECTION 10

Settings

Drafting Settings	10-2
Drawing Limits	10-3
Linetype Scale	10-4
Lineweights	10-5
Pick box	10-6
Units/Precision	10-7

SECTION 11

Text

Creating a new Text Style	11-2
Changing a Text Style	11-4
Multiline Text	11-5
MTJIGSTRING	11-7
Multiline Text Line Spacing	11-8
Editing Multiline Text	11-9
Indents	11-10
Tabs	11-11
Scaling Text	11-12
Single Line Text	11-13
Editing Single Line text	11-15
Special Text Characters	11-16

SECTIONS 12

UCS

Displaying the UCS icon	12-2
Moving the Origin	12-3

Section 1
Action Commands

ARRAY

The ARRAY command allows you to make multiple copies in a **RECTANGULAR** or Circular **(POLAR)** pattern. The maximum limit of copies per array is 100,000. This limit can be changed but should accommodate most users.

RECTANGULAR ARRAY
The method allows you to make multiple copies of object(s) in a rectangular pattern. You specify the number of rows (horizontal), columns (vertical) and the offset distance between the rows and columns. The offset distances will be equally spaced.

Offset Distance is sometimes tricky to understand. *Read this carefully*. The offset distance is the distance from a specific location on the original to that same location on the invisible copy. It is not just the space in between the two. Refer to the example below.

To use the rectangular array command you will select the object(s), specify how many rows and columns desired and the offset distance for the rows and the columns. **Step by step instructions on the following pages.**

Example of Rectangular Array:

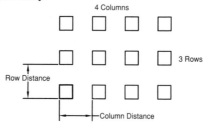

Example of Rectangular Array on an angle:
Notice the copies do not rotate. The Angle is only used to establish the placement.

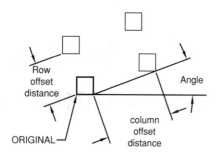

ARRAY (Continued)

RECTANGULAR ARRAY

1. Select the ARRAY command using one of the following:

> **TYPE = ARRAY**
> **PULLDOWN = MODIFY / ARRAY**
> **TOOLBAR = MODIFY**

The dialog box shown below will appear.

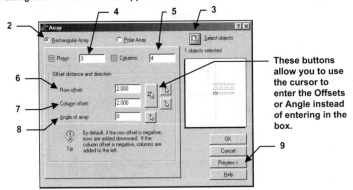

These buttons allow you to use the cursor to enter the Offsets or Angle instead of entering in the box.

2. Select "**Rectangular Array**".

3. Select the **"Select Objects"** button.
   This will take you back to your drawing. Select the objects to Array then <enter>.

4. Enter the number of rows.

5. Enter the number of columns.

6. Enter the row offset. (The distance from a specific location on the original to that same specific location on the future copy.) See example on previous page.

7. Enter the column offset. (The distance from a specific location on the original to that same specific location on the future copy.) See example on previous page.

8. Enter an angle if you would like the array to be on an angle.

9. Select the Preview button.
   If it looks correct, select the Accept button. If it is not correct, select the Modify button, make the necessary corrections and preview again.
   (Note: If the Preview button is gray, you have forgotten to "Select objects"- #3)

ARRAY (continued)

POLAR ARRAY

This method allows you to make multiple copies in a circular pattern. You specify the total number of copies to fill a specific Angle or specify the angle between each copy and angle to fill.

To use the polar array command you select the object(s) to copy, specify the center of the array, specify the number of copies or the angle between the copies, the angle to fill and if you would like the copies to rotate as they are copied.

Example of Polar Array

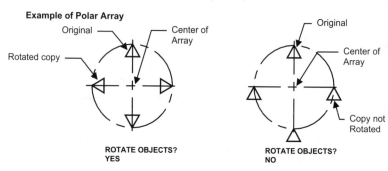

ROTATE OBJECTS?
YES

ROTATE OBJECTS?
NO

Note: the two examples above use the objects default base point. The example below specifies the base point for the copies. The end result is different from the example above on the right.

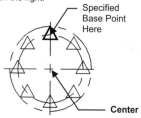

ROTATE OBJECTS?
NO
(Specified base point)

DEFAULT OBJECT BASE POINTS	
Type of Object	**Default Base Pt.**
Arc, Circle, Ellipse	Center
Polygon, Rectangle	First Corner
Line, Polyline, Donut	Starting Point
Text, Block	Insertion Point

Note: If you select multiple objects the base Point of the last object selected is used to contruct the array.

The difference between "Center Point" and "Object Base Point" is sometimes confusing. When specifying the Center Point, try to visualize the copies already there. Now, in your mind, try to visualize the center of that array. In other words, it is the Pivot Point from which the copies will be placed around. The Base Point is different. It is located on the original object.

ARRAY (continued)

POLAR ARRAY

1. Select the ARRAY command using one of the following:

> **TYPE = ARRAY**
> **PULLDOWN = MODIFY / ARRAY**
> **TOOLBAR = MODIFY**

The dialog box shown below will appear.

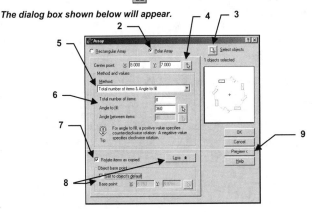

2. Select **" Polar Array".**

3. Select the **"Select Objects"** button.
   This will take you back to your drawing. Select the objects to Array then <enter>.

4. Press the "Center Point" button and select the center point with the cursor or enter the X and Y coordinates in the "Center Point" boxes.

5. Select the method.

6. Enter the "Total number of items", "Angle to fill" or "Angle between items".

7. Select whether you want the items rotated as copied or not.

8. Accept the object's default base point or enter the X and Y coordinates.
   (Select the "More" button to show this area. Select the "Less" button to not show.)

9. Select the Preview button.
   If it looks correct, select the Accept button. If it is not correct, select the Modify button, make the necessary corrections and preview again.
   (Note: If the Preview button is gray, you have forgotten to select objects #3)

BREAK

The **BREAK** command allows you to break a space in an object, break the end off an object or split a line in two. I think of it as taking a bite out of an object.
There are 4 break methods, as described below.

You may select the **BREAK** command by using one of the following:

> TYPING = BR
> PULLDOWN = MODIFY / BREAK
> TOOLBAR = MODIFY

METHOD 1
How to break one object into two separate objects with no visible space in between. (Use this method if the location of the break is not important)

a. Draw a line.
b. Select the **BREAK** command using one of the methods listed above.
c. _break Select object: *pick the break location (P1) by clicking on it.*
d. Specify second break point or [First point]: *type @ <enter>.*
(This will duplicate the last point.)
e. Now if you click on one end of the line you will see that there are 2 lines instead of just one.

METHOD 2
This method is the same as method 1, however, **use this method if the location of the break is very specific.**

a. Select the **BREAK** command using one of the methods listed above.
b. _break Select objects: *select the object to break (P1).*
c. Specify second break point or [First point]: *type F <enter>.*
d. Specify first break point: *select break location (P2) accurately.*
e. Specify second break point: *type @ <enter>.*

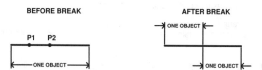

**Method 2 can also be accomplished easily by selecting the "Break at point"
icon.   But I wanted you to understand how it works.**

BREAK (continued)

<u>METHOD 3</u>
Take a bite out of an object. (Use this method if the location of the BREAK is not important.)
(This is the Default option.)

a. Select the **BREAK** command.
b. _break Select object: *pick the first break location (P1).*
c. Specify second break point or [First point]: *pick the second break location (P2).*

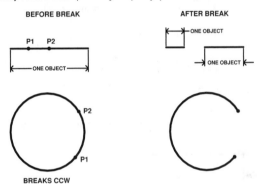

Note: Circles can't be broken with "1 point". You must use 2 points.

<u>METHOD 4</u>
This method is the same as method 3, however, <u>use this method if the location of the break is very specific.</u>

a. Select the **BREAK** command.
b. _break Select objects: *select the object to break (P1) anywhere on the object.*
c. Specify second break point or [First point]: *type F <enter>.*
d. Specify first break point: *select the first break location (P2) accurately.*
e. Specify second break point: *select the second break location (P3) accurately.*

CHAMFER

The **CHAMFER** command allows you to create a chamfered corner on two lines. There are two methods: **Distance and Angle**.

DISTANCE METHOD
1. Select the CHAMFER command using one of the following:

> **TYPE = CHA**
> **PULLDOWN = MODIFY / CHAMFER**
> **TOOLBAR = MODIFY**

(**Distance Method** requires input of a distance for each side of the corner.)

Command: _chamfer
(TRIM mode) Current chamfer Dist1 = 0.000, Dist2 = 0.000
Select first line or [Undo/Polyline/Distance/Angle/Trim/mEthod/Multiple]: *select "D"<enter>.*
Specify first chamfer distance <0.000>: *type the distance <enter>.*
Specify second chamfer distance <1.000>: *type the distance <enter>.*

2. **NOW CHAMFER THE OBJECT**
Select first line or [Undo/Polyline/Distance/Angle/Trim/mEthod/Multiple]:
select the (First Line) to be chamfered (dist 1 side).
Select second line or shift-select to apply corner: *select the (Second Line) to be chamfered (dist 2 side).*

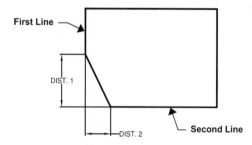

Polyline: This option allows you to Chamfer all intersections of a Polyline in one operation. Such as all 4 corners of a rectangle.

Trim: This option controls whether the original lines are trimmed or remain after the corners are chamfered. (Set to Trim or No trim.)

mEthod: Allows you to switch between **Distance** and **Angle** method. The distance or angle must have been set previously.

Multiple: Repeats the Chamfer command until you press <enter> or esc key.

CHAMFER (continued)

ANGLE METHOD
1. Select the CHAMFER command

   (**Angle method** requires input for the length of the line and an angle)

   Command: _chamfer
   (TRIM mode) Current chamfer Dist1 = 1.000, Dist2 = 1.000
   Select first line or [Undo/Polyline/Distance/Angle/Trim/method/Multiple]:
   select "A" <enter>
   Specify chamfer length on the first line <0.000>: *type the chamfer length <enter>*
   (dist 1)
   Specify chamfer angle from the first line <0>: *type the angle <enter>*

2. **NOW CHAMFER THE OBJECT**
   Select first line or [Undo/Polyline/Distance/Angle/Trim/mEthod/Multiple]: *select the (First Line) to be chamfered. (the length side)*
   Select second line or shift-select to apply corner: *select the (second line) to be chamfered. (the Angle side)*

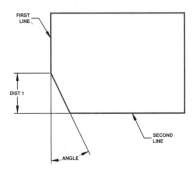

COPY

The COPY command creates a duplicate set of the objects selected.   The COPY command is similar to the MOVE command. You must select the objects to be copied, select a base point and a new location. The difference is, the Move command merely moves the objects to a new location. The Copy command makes a copy and you select the location for the new copy.

Select the Copy command using one of the following commands:

TYPE = CO
PULLDOWN = MODIFY / COPY
TOOLBAR = MODIFY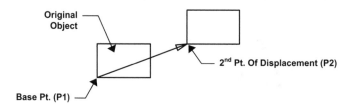

Command: _copy
Select objects: *select the objects you want to copy*
Select objects: *stop selecting objects by selecting <enter>*
Specify base point or displacement: *select a base point (P1) (usually on the object)*
Specify second point of displacement or <use first point as displacement>: *select the new location (P2) for the first copy*
Specify second point or [Exit / Undo] <Exit>: *select the new location (P2) for the next copy or press <enter> to exit.*

Original
Object

2nd Pt. Of Displacement (P2)

Base Pt. (P1)

DIVIDE

The DIVIDE command divides an object mathematically by the NUMBER of segments you designate. It then places a POINT (object) at each interval on the object.

Note: the object selected is **NOT** broken into segments. The POINTS are simply drawn **ON** the object.

First select the **POINT STYLE** to be placed on the object.

Next select the **DIVIDE** command using one of the following:

> **TYPE = DIV**
> **PULL DOWN = DRAW / POINT / DIVIDE**
> **TOOLBAR = DRAW**

Select object to divide: *select the object to divide.*
Enter the number of segments or [Block]: *type the number of segments <enter>*

EXAMPLE:

This LINE has been DIVIDED into 4 EQUAL lengths.
But remember, the line is not broken into segments.
The Points are simply drawn ON the object.

ERASE

There are 3 methods to **erase** (delete) objects from the drawing.
You decide which one you prefer to use. They all work equally well.

METHOD 1.
Select the Erase command first and then select the objects.

1. Start the Erase command by using one of the following:

     TYPING = E <enter>
     PULLDOWN = MODIFY / ERASE
     TOOLBAR = MODIFY 📝

2. Select objects: *pick one or more objects*
   Select objects: *press <enter> and the objects will disappear*

METHOD 2.
Select the Objects first and then the <u>Erase</u> command from the shortcut menu.

1. Select the object(s) to be erased.
2. Press the right mouse button.
3. Select **Erase** from the short-cut menu.

METHOD 3.
Select the Objects first and then the <u>Delete</u> key

1. Select the object(s) to be erased.
2. Press the <u>Delete</u> key.

NOTE: Very important
If you want the erased objects to return, press **U <enter>** or **Ctrl + Z** or
the **Undo arrow icon.** ↩

This will **"Undo"** the effects of the last command.

Read more about the **Undo** and **Redo** commands.

EXPLODE

The EXPLODE command changes (explodes) an object into its primitive objects.
For example: A rectangle is originally one object. If you explode it, it changes into 4
lines. Visually you will not be able to see the change unless you select one of the lines.

1. Select the Explode command by using one of the following:

   TYPE =X
   PULLDOWN = MODIFY / EXPLODE
   TOOLBAR = MODIFY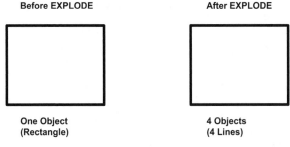

2. The following will appear on the command line:

Command: _explode
Select objects: *select the object(s) you want to explode.*
Select objects: *select <enter>.*

Before EXPLODE	After EXPLODE

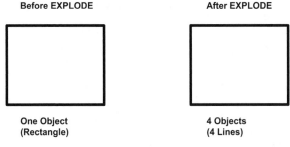

One Object   **4 Objects**
(Rectangle)   **(4 Lines)**

**(Notice there is no visible difference. But now you have 4 lines instead of 1
Rectangle)**

Try this:
Draw a rectangle and then click on it. The entire object highlights.
Now explode the rectangle, then click on it again. Only the line you clicked on should
be highlighted. Each line that forms the rectangular shape is now an individual object.

EXTEND

The **EXTEND** command is used to extend an object to a **boundary.** The object to be extended must actually or theoretically intersect the boundary.

1. Select the **EXTEND** command using one of the following:

   TYPE = EX
   PULLDOWN = MODIFY / EXTEND
   TOOLBAR = MODIFY [⊟]

2. The following will appear on the command line:

Command: _extend
Current settings: Projection = UCS   Edge = Extend
Select boundary edges ...
Select objects or <select all>: **select boundary (P1*below*) by clicking on the object.**
Select objects: **stop selecting boundaries by selecting <enter>.**
Select object to extend or shift-select to Trim or
[Fence/Crossing/Project/Edge/Undo]:**select the object that you want to extend (P2 and P3). (Select the end of the object that you want to extend.)**
Select object to extend or [Fence/Crossing/Project/Edge/Undo]:**stop selecting objects by selecting <enter>.**

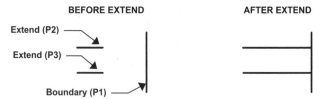

| BEFORE EXTEND | AFTER EXTEND |

Extend (P2)

Extend (P3)

Boundary (P1)

You may toggle between Extend and Trim. Hold down the shift key and the Trim command is activated. Release the shift key and you return to Extend.

Project   Same as Edge except used only in "3D".
Edge   (Extend or No Extend)
In the **"Extend"** mode, (default mode) the boundary and the Objects to be extended need only imaginarily intersect if the objects were infinite in length.
In the **"No Extend"** mode the boundary and the objects to be extended must visibly intersect.
Crossing   You may select objects using a Crossing Window.
eRase   You may erase an object instead of trimming while in the Trim command.
Undo   You may "undo" the last trimmed object while in the Trim command.

FILLET

The FILLET command will create a radius between two objects. The objects do not have to be touching. If two parallel lines are selected, it will construct a full radius.

RADIUS A CORNER

1. Select the FILLET command using one of the following:

TYPE = F
PULLDOWN = MODIFY / FILLET
TOOLBAR = MODIFY

The following will appear on the command line:

2. SET THE RADIUS OF THE FILLET
Command: _fillet
Current settings: Mode = TRIM, Radius = 0.000
Select first object or [Undo/Polyline/Radius/Trim/Multiple]: *type "R" <enter>*
Specify fillet radius <0.000>: *type the radius <enter>*

3. NOW FILLET THE OBJECTS
Select first object or [Undo/Polyline/Radius/Trim/Multiple]: *select the first object to be*
filleted
Select second object or shift-select to apply corner: *select the second object to be*
filleted

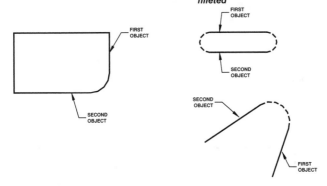

Polyline: This option allows you to fillet all intersections of a Polyline in one operation, such as all 4 corners of a rectangle.

Trim: This option controls whether the original lines are trimmed to the end of the Arc or remain the original length. (Set to Trim or No trim)

Multiple: Repeats the fillet command until you press <enter> or esc key.

FILLET continued...

The FILLET command may also be used to create a square corner.

SQUARE CORNER

1. Select the FILLET command using one of the following:

    TYPE = F
    PULLDOWN = MODIFY / FILLET
    TOOLBAR = MODIFY

The following will appear on the command line:

2. Select the two lines to form the square corner
Select first object or [Undo/Polyline/Radius/Trim/Multiple]: *select the first object (P1)*

Select second object or shift-select to apply corner: ***Hold the shift key down while selecting the second object (P2)***

ORIGINAL

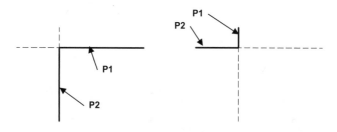

Note: The corner direction depends on which end of the object you select.

INQUIRY

The INQUIRY command allows you to Inquire about objects on the screen. There are 8 options within the inquiry menu but only the 3 most commonly used are discussed here. Refer the AutoCAD's "Help" menu for the remaining 5.

LIST

The LIST command will list the type of object you have selected, coordinate location and properties that apply to the object.

1. Select the LIST command using one of the following:

TYPE = LIST
PULLDOWN = TOOLS / INQUIRY / List
TOOLBAR = STANDARD

2. Select the object: *select the object*
3. Select the object: *press <enter> to stop*
(The information will be listed in the "Text Screen". Press F2 to close the Text Screen.)

DISTANCE

The DISTANCE command will list the distance between two points that you select.

1. Select the DISTANCE command using one of the following:

TYPE = DI
PULLDOWN = TOOLS / INQUIRY / Distance
TOOLBAR = STANDARD

2. First point: *select the first point*
3. Second point: *select the second point*
    Distance = *distance between the two points will be listed here*

ID POINT

The ID POINT or LOCATE POINT command will list the X and Y coordinates of the point that you selected. The coordinates will be from the ORIGIN.

1. Select the ID POINT command

TYPE = ID
PULLDOWN = TOOLS / INQUIRY / ID Point
TOOLBAR = TOOLS

2. Point: *select a point anywhere on the drawing*
    X =*coordinate listed here* Y = *coordinate listed here* Z = *coordinate listed here*

Note: the **ID POINT** can also be used to create a ***LAST POINT***. This enables you to use Relative coordinates (the @ symbol) for the location of the next object.

MATCH PROPERTIES

Match Properties is used to "paint" the properties of one object to another. This is a simple and useful command. You first select the object that has the desired properties (the source object) and then select the object you want to "paint" the properties to (destination object).

Only one "source object" can be selected but its properties can be painted to any number of "destination objects".

1. Select the Match Properties command using one of the following:

> **TYPE = MATCHPROP or MA**
> **PULLDOWN = MODIFY / MATCH PROPERTIES**
> **TOOLBAR = STANDARD**

Command: matchprop

2. Select source object: *select the object with the desired properties to match*

3. Select destination object(s) or [Settings]: *select the object(s) you want to receive the matching properties.*

4. Select destination object(s) or [Settings]: *select more objects or <enter> to stop.*

Note: If you do not want to match all of the properties, right click and select "Settings" from the short cut menu, before selecting the destination object. Uncheck all the properties you do not want to match and select the OK button. Then select the destination object(s).

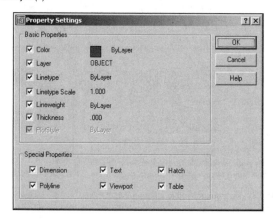

MEASURE

The **MEASURE** command is very similar to the **DIVIDE** command because point objects are drawn at intervals on an object. However, the **MEASURE** command allows you to designate the **LENGTH** of the segments rather than the number of segments.

Note: the object selected is **NOT** broken into segments. The **POINTS** are simply drawn **ON** the object.

First select the **POINT STYLE** to be placed on the object.

Next select the **MEASURE** command using one of the following:

> **TYPE = ME**
> **PULL DOWN = DRAW / POINT / MEASURE**
> **TOOLBAR = DRAW** ⟋

Command: _measure
Select object to measure: *select the object to measure.*
(Note: this selection point is also where the MEASUREment will start.)
Specify length of segment or [Block]: *type the length of one segment <enter>*

EXAMPLE:

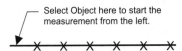

Select Object here to start the measurement from the left.

The MEASUREment was started at the left endpoint, and ended just short of the right end of the line. The remainder is less than the measurement length designated.

MIRROR

The MIRROR command allows you to make a mirrored image of any objects you select.
You can use this command for creating right / left hand parts. You can draw a
symmetrical object more efficiently by only drawing half of it.

Select the **MIRROR** command using one of the following:

> **TYPE = MI**
> **PULLDOWN = MODIFY / MIRROR**
> **TOOLBARS = MODIFY**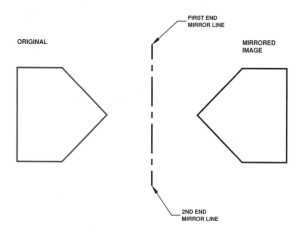

The following will appear on the command line:

Command: _mirror
Select objects: ***select the objects to be mirrored***
Select objects: ***stop selecting objects by selecting \<enter>***
Specify first point of mirror line: ***select the first end of the mirror line***
Specify second point of mirror line: ***select the second end of the mirror line***
Erase source objects? [Yes/No] \<N>: ***select Y or N***

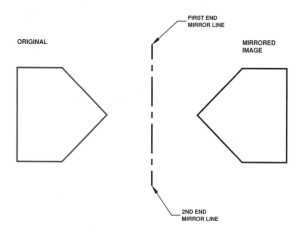

How to control text when using the Mirror command:
1. At the command line type   **mirrtext \<enter>**
2. Select the setting by typing 0 or 1 \<enter>.
0 = Retains text direction   1 = Mirrors the text (default setting)

| MIRRTEXT SETTING 0 | MIRRTEXT SETTING 0 |
| MIRRTEXT SETTING 1 | 1 ƃNITTƎS TXƎTЯЯIM |

MOVE

The MOVE command is used to move object(s) from their current location (basepoint) to a new location (second displacement point).

1. Select the Move command using one of the following:

   TYPE = M
   PULLDOWN = MODIFY / MOVE
   TOOLBAR = MODIFY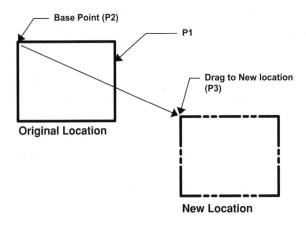

2. The following will appear on the command line:

   Command: _move
   Select objects: *select the object(s) you want to move (P1).*
   Select objects: *stop selecting object(s) by selecting <enter>.*
   Specify base point or displacement: *select a location (P2) (usually on the object).*
   Specify second point of displacement or <use first point as displacement>: *move the object to it's new location (P3) and left click.*

Note: If you press <enter> instead of actually picking a new location (P3), Autocad will send it into <u>Outer Space</u>. If this happens, press U <enter> or select the " undo" icon *and try again.*

Base Point (P2)

P1

Drag to New location (P3)

Original Location

New Location

OFFSET

The **OFFSET** command duplicates an object parallel to the original object at a specified distance. You can offset Lines, Arcs, Circles, Ellipses, 2D Polylines and Splines.

EXAMPLES:

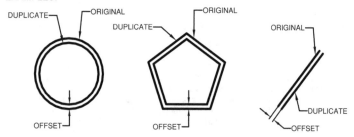

How to use the OFFSET command:

1. Select the OFFSET command using one of the following:

       TYPING = OFFSET
       PULLDOWN = MODIFY / OFFSET
       TOOLBAR = MODIFY

2. Specify offset distance or [Through/Erase/Layer] <Through>: *type the offset distance or select Erase or Layer. (see option on the next page)*

3. Select object to offset or <Exit/Undo>: *select the object to offset.*

4. Specify point on side to offset or [Exit/Multiple/Undo]<Exit>: *Select which side of the original you want the duplicate to appear by placing your cursor and clicking. (See options on the next page)*

5. Select object to offset or [Exit/Undo]<Exit>: *Press <enter> to stop.*

OPTIONS:

Through: Creates an object passing through a specified point.

Erase: Erases the source object after it is offset.

Layer: Determines whether offset objects are created on the <u>current</u> layer or on the layer of the <u>source</u> object. Select <u>Layer</u> and then select <u>current</u> or <u>source.</u>

Multiple: Turns on the multiple offset mode, which allows you to continue creating duplicates of the original without re-selecting the original.

Exit: Exits the Offset command.

Undo: Reverses the previous offset.

OFFSETGAPTYPE (Version 2006 only and not available in 2006 LT)

When you offset a closed 2D object, such as a rectangle, to create a **<u>larger</u>** object it results in potential gaps between the segments. The **offsetgaptype** system variable controls how these gaps are closed.

To set the offsetgaptype:

Type: *offsetgaptype <enter>*

Enter one of the following:

      0 = Fills the gap by extending the polyline segments. (default setting)

      1 = Fills the gap with <u>filleted arc segments</u>. The radius of each arc segment is equal to the offset distance.

      2 = Fills the gap with <u>chamfered line segments.</u> The perpendicular distance to each chamfer is equal to the offset distance.

ROTATE

The **ROTATE** command is used to rotate objects around a Base Point. (pivot point)
After selecting the objects and the base point, you will enter the rotation angle or select a
reference angle followed by the new angle.
A **Positive** rotation angle revolves the objects **Counter- Clockwise**.
A **Negative** rotation angle revolves the objects **Clockwise**.

Select the ROTATE command using one of the following:

> **TYPE =RO**
> **PULLDOWN = MODIFY / ROTATE**
> **TOOLBAR = MODIFY**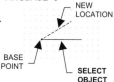

ROTATION ANGLE OPTION
  Command: _rotate
1. Current positive angle in UCS: ANGDIR=counterclockwise ANGBASE=0
   Select objects: *select the object to rotate.*
2. Select objects: *select more object(s) or <enter> to stop.*
3. Specify base point: *select the base point (pivot point).*
4. Specify rotation angle or [Copy/Reference]<0>:
> *type the angle of rotation.*

REFERENCE OPTION
  Command: _rotate
1. Current positive angle in UCS: ANGDIR=counterclockwise ANGBASE=0
   Select objects: *select the object to rotate.*
2. Select objects: *select more object(s) or <enter> to stop.*
3. Specify base point: *select the base point (pivot point).*
4. Specify rotation angle or [Reference]: *select Reference.*
5. Specify the reference angle <0>: *Snap to the reference object (1) and (2).*
6. Specify the new angle: *drag the object and snap to the new angle.*

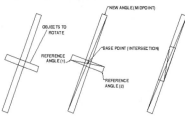

COPY OPTION
Creates a duplicate of the selected object(s). The duplicate is directly on top of the
original. The duplicate will be rotated. The Original remains the same.

SCALE

The **SCALE** command is used to make objects larger or smaller <u>proportionately</u>. You may scale using a scale factor or a reference length. You must also specify a base point. Think of the base point as a stationary point from which the objects scale. It does not move.

1. Select the SCALE command using one of the following:

> **TYPE = SCALE**
> **PULLDOWN = MODIFY / SCALE**
> **TOOLBAR = MODIFY**

SCALE FACTOR
> Command: _scale
2. Select objects: ***select the object(s) to be scaled***
3. Select objects: ***select more object(s) or <enter> to stop***
4. Specify base point: ***select the stationary point on the object***
5. Specify scale factor or [Copy/Reference]: ***type the <u>scale factor</u> <enter>***

If the scale factor is greater than 1, the objects will increase in size.
If the scale factor is less than 1, the objects will decrease in size.

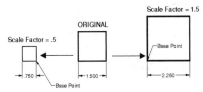

REFERENCE
Command: _scale
2. Select objects: ***select the object(s) to be scaled***
3. Select objects: ***select more object(s) or <enter> to stop***
4. Specify base point: ***select the stationary point on the object***
5. Specify scale factor or [Copy/Reference]: ***select Reference***
6. Specify reference length <1>: ***specify a <u>reference</u> length***
7. Specify new length: ***specify the new length***

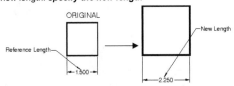

COPY creates a duplicate of the selected object. The duplicate is directly on top of the original. The duplicate will be scaled. The Original remains the same.

STRETCH

The **STRETCH** command allows you to stretch or compress object(s). Unlike the Scale command, you can alter an objects proportion with the Stretch command. In other words, you may increase the length without changing the width and vice versa.

Stretch is a very valuable tool. Take some time to really understand this command. It will save you hours when making corrections to drawings.

When selecting the object(s) you must use a **CROSSING** window.
Objects that are crossed, will **stretch.**
Objects that are totally enclosed, will **move**.

1. Select the STRETCH command using one of the following:

   TYPE = S
   PULLDOWN = MODIFY / STRETCH
   TOOLBAR = MODIFY

   Command: _stretch
2. Select objects to stretch by crossing-window or crossing-polygon...
   Select objects: **select the first corner of the crossing window**
3. Specify opposite corner: **specify the opposite corner of the crossing window**
4. Select objects: **<enter>**
5. Specify base point or [Displacement] <Displacement>: **select a base point (where it stretches from)**
6. Specify second point or <use first point as displacement>: **type coordinates or place location with cursor**

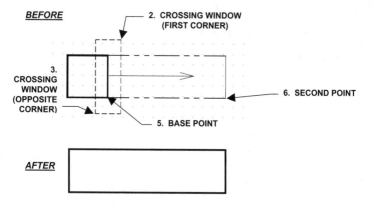

<u>BEFORE</u>

2. **CROSSING WINDOW (FIRST CORNER)**

3. **CROSSING WINDOW (OPPOSITE CORNER)**

5. **BASE POINT**

6. **SECOND POINT**

<u>AFTER</u>

TRIM

The **TRIM** command is used to trim an object to a **cutting edge**. You first select the "Cutting Edge" and then select the part of the object you want to trim. The object to be trimmed must actually intersect the cutting edge or could intersect if the objects were infinite in length.

1. Select the Trim command using one of the following:

   TYPE = TR
   PULLDOWN = MODIFY / TRIM
   TOOLBAR = MODIFY

2. The following will appear on the command line:

Command: _trim
Current settings: Projection = UCS   Edge = Extend
Select cutting edges ...
Select objects or <select all>: *select cutting edge(s) by clicking on the object (P1)*
Select objects: *stop selecting cutting edges by pressing the <enter> key*
Select object to trim or shift-select to extend or
[Fence/Crossing/Project/Edge/eRase/Undo]: *select the object that you want to trim. (P2)*
(Select the part of the object that you want to disappear, not the part you want to remain)
Select object to trim or [Fence/Crossing/Project/Edge/eRase/Undo]: *press <enter> to stop*

Note: You may toggle between Trim and Extend. Hold down the shift key and the
      Extend command is activated. Release the shift key and you return to Trim.

BEFORE TRIM   **AFTER TRIM**

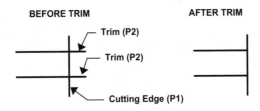

Project   Same as Edge except used only in "3D".
Edge   (Extend or No Extend)
In the **"Extend"** mode, (default mode) the boundary and the Objects to be extended need only imaginarily intersect if the objects were infinite in length.
In the **"No Extend"** mode the boundary and the objects to be extended must visibly intersect.
Crossing   You may select objects using a Crossing Window.
eRase   You may erase an object instead of trimming while in the Trim command.
Undo   You may "undo" the last trimmed object while in the Trim command.

UNDO and REDO

The **UNDO** command allows you to undo <u>previous commands</u>. For example, if you erase an object by mistake, you can UNDO the previous "erase" command and the object will reappear. So don't panic if you do something wrong. Just use the UNDO command to remove previous commands.

Note:
You may UNDO commands used during a work session until you close the drawing.

How to use the "Undo" command.

1. Start a new drawing.
2. Draw a line, rectangle and a circle.

Your drawing should look approximately like this.

3. Erase the Rectangle and the Circle.

(The Circle and the Rectangle disappear.)

4. Select the UNDO down arrow and select the ERASE command from the list of commands.

You have now deleted the ERASE command operation. As a result the erased objects reappear.

How to use the Redo command:

Select the REDO down arrow and you will see that the ERASE command moved to the REDO list. If you would like to bring the ERASE operation back, select the ERASE command from the list. The Circle and Rectangle will disappear again.

WIPEOUT

The Wipeout command creates a blank area that covers existing objects. The area has a background that matches the background of the drawing area. This area is bounded by the wipeout frame, which you can turn on or off.

1. Select the Wipeout command using one of the following:

        TYPE = WIPEOUT
        PULLDOWN = DRAW / WIPEOUT
        TOOLBAR = NONE

2. Command: _wipeout Specify first point or [Frames/Polyline] <Polyline>: *specify the first point of the shape (P1)*
3. Specify next point: *specify the next point (P2)*
4. Specify next point or [Undo]: *specify the next point (P3)*
5. Specify next point or [Undo]: *specify the next point (P4)*
6. Specify next point or [Close/Undo]: *specify the next point or <enter> to close*

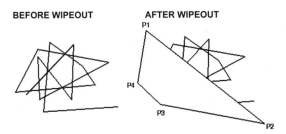

BEFORE WIPEOUT   **AFTER WIPEOUT**

TURNING FRAMES ON OR OFF

1. Select the Wipeout command.
2. Select the "Frames" option.
3. Enter ON or OFF.

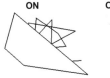

ON   **OFF**

Note: If you want to move the objects and the wipeout area, move them together not separately. If you do move the objects and the wipeout, the objects under the wipeout area may reappear. Select **View / Regen** and they will disappear again.

ZOOM

The **ZOOM** command is used to move closer or farther away to an object.

The following is an example of **Zoom / Window** to zoom in closer to an object.

1. Select the Zoom command by using one of the following:

    TYPING = Z <enter>
    PULLDOWN = VIEW / ZOOM
    TOOLBAR = STANDARD

2. Select the **"Window"** option and draw a window around the area you wish to magnify by moving the cursor to the lower left area of the object(s) and left click. Then move the cursor diagonally to form a square shape around the objects, and left click again. *(Do not hold the left mouse button down while moving the cursor, just click it at the first and diagonal corners of the square shape)*

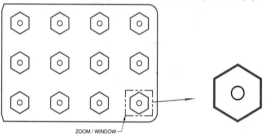

ZOOM / WINDOW

Additional Zoom options described below. (Try them)

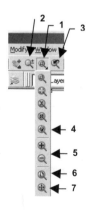

1. **WINDOW** = zoom in on an area by specifying a window (rectangle) around the area.

2. **REAL TIME** = Interactive Zoom. You can zoom in or out by moving the cursor vertically up or down while pressing the left mouse button. To stop, press the Esc key.

3. **PREVIOUS** = returns the screen to the previous display. (Limited to 10 previous displays)

4. **OBJECT** = zooms in on a selected object

5. **IN or OUT** = moves in 2X or out 2X

6. **All** = Changes the screen to the size of the drawing limits. If you have objects outside of the drawing limits, Zoom/All will display them too.

7. **EXTENTS** = Displays all objects in the drawing file, using the smallest window possible.

Section 2
Concepts

MODEL and LAYOUT tabs

Read this information carefully. It is very important that you understand this concept.

AutoCAD provides two drawing spaces, **MODEL** and **LAYOUT**. You move into one or the other by selecting either the MODEL or LAYOUT tabs, located at the bottom left of the drawing area.

<u>Model Tab</u> (Also called *Model Space*)
When you select the Model tab you enter <u>MODEL SPACE</u>.
Model Space is where you **create** and **modify** your drawings.

<u>Layout1 Tab</u> (Also called *Paper Space*)

When you select a Layout tab you enter <u>PAPER SPACE</u>.
The primary function of Paper Space is to prepare the drawing for plotting.

When you select the Layout tab for the first time, the "<u>Page Setup Manager" dialog box</u> will appear. Using this dialog box, you will assign a name for your new page setup. Then you will specify which plotting device and paper size to use for plotting. For now, select the CLOSE button to continue.
(More information on this in "How to create a Page Setup")

Notice that Model Space seems to have disappeared, and a <u>blank sheet of paper</u> is displayed on the screen. This sheet of paper is basically in front of the Model Space. (See illustration below) To see your drawing (Model Space) while still in Paper Space, you must <u>cut a hole</u> in this sheet. This hole is called a **"Viewport"**. *(Refer to "Viewports")*

<u>Try to think of this as a picture frame (paper space) in front of a photograph (model space).</u>

Generally, the only objects that should be in paper space are the Title Block, Border, Dimensions and Notes.

Viewport

Model Space:
The Drawing should be drawn and modified here.

Paper Space:
Title Block, Border, Dimensions and Notes should be drawn here

See through Viewport to Model Space

Click on Model tab to enter Model Space.

Click on Layout1 tab enter Paper Space.

MODEL SPACE

PAPER SPACE

CREATING SCALED DRAWINGS

A very important rule in CAD you must understand is: *"All objects are drawn full size".* In other words, if you want to draw a line 20 feet long, you actually draw it 20 feet long. If the line is 1/8" long, you actually draw it 1/8" long.

Drawing and Plotting objects that are very large or very small.

In the previous lessons you created medium sized drawings. Not too big, not too small. But what if you wanted to draw a house? Could you print it to scale on a 17 X 11 piece of paper? How about a small paper clip. Could you make it big enough to dimension? Let's start with the house.

Drawing something <u>large</u> such as a house.
1. Start a new drawing from scratch.
2. Set the units for the drawing to Architectural.
3. Set the snap to 3 inches and set the grids to 1 foot.
4. Set the drawing limits, in model space, to:
   Lower left corner: 0, 0 Upper Right Corner: 45', 35' (feet), then (Zoom / All).
   Now your drawing area is big enough for the entire house to be drawn <u>full size</u>.
5. Draw a rectangle: 30' L X 20' W (representing the house) and draw a pretend roof.

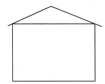

Plotting the house drawing
Now you want to plot this house drawing on a 17 x 11 piece of paper. This is where paper space makes it easy.

6. Select the Layout tab (paper space).
7. When the Page Setup Manager appears, select NEW, give it a name, then OK.
8. Specify the plotter LaserJet 4MV, the paper size 11 x 17 then OK.
9. Select "Set Current" and "Close".
10 Open the "Viewport" tool bar and cut a viewport so you can see through the
    Viewport to the house drawing that is in model space.
11 Double click inside the Viewport to reach through to model space.
12 Select <u>View / Zoom / Extents</u> so the entire house is visible within the viewport frame.

Now here is where the magic happens.

13. Adjust the scale of model space.
    Scroll down the list of scales and select ¼" = 1'.

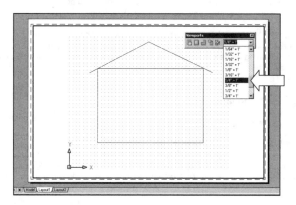

Wow! A 30ft x 20ft house fit on to an 11" x 17" sheet of paper.

Let's talk about what really happened.
Try to picture yourself standing in front of your house with an empty picture frame in your hands. Look at your house through the picture frame. The house is way too big to fit in the frame. So you walk across the street and look through the picture frame again. Does the house appear smaller? If you could walk far enough away from the house it would eventually appear small enough to fit in the picture frame. But....the house did not actually change size, did it? It only appears smaller because you are farther away from it.

This is the concept that you need to understand.

When you adjust the scale of model space it does not change the actual size of the objects. They just appear smaller because they are farther away from paper space. When you dimension the house, the dimension values will be the actual measurement of the house. In other words, the 30 ft. width will have a dimension of 30'-0".

14. Lock the viewport.
    When you have adjusted the model space scale to your satisfaction you should "lock" the viewport so the scale will not change when you zoom.
15. Save this drawing as "HOUSE".
16. Plot the paper space / model space combination using File / Plot.
    The Page Setup scale should remain 1 : 1 (Remember you have already adjusted the model space scale. You do not need to scale the combination it is just fine.)
17. Don't forget to Preview before plotting.

Drawing something <u>small</u> such as a paperclip.

When plotting something smaller you have to move the picture frame closer to model space rather than farther away. So in that case, you would adjust the model space scale to a scale larger than 1 : 1 until the object in model space appears large enough to see easily and dimension. Remember, even though the object appears larger, when you dimension it the dimension values will be the correct information.

Let's try a small object.

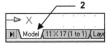

1. Open **My Decimal Setup.**
2. Select the "Model" tab.
3. Draw a Rectangle 2" L X 1" W. (We won't actually draw the paperclip right now.)

4. Select the "**11 x 17 (1 to 1)** " tab.

You should now be able to see the Rectangle in the viewport.

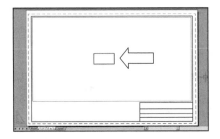

5. Unlock the Viewport.
6. Double click inside the Viewport to get into model space.
7. Adjust the scale of model space to 4 : 1 .

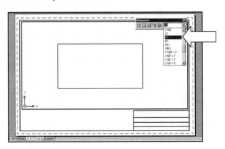

Now the Rectangle is large enough to see clearly and dimension.

Section 3
Dimensioning

DIMENSIONING

True Associative
True Associative Dimensioning means that the dimensions are actually attached to the objects that they dimension. If you move the object, the dimension will move with it. If you scale or stretch the object, the dimension text value will change also. (Note: It is not parametric. This means, you cannot change the dimension text value and expect the object to change.)

True Associative Dimensioning can be set to ON, OFF or Exploded.
I strongly suggest that you keep true associative dimensioning on. It is truly a very powerful feature and will make editing the objects and dimensions much easier.

> **On** = Dimensions are truly associative. The dimensions are associated to the objects and will change if the object is changed. (Setting 2)

> **Off** = Dimensions are non-associative. The dimensions are not associated to the objects and will not change if the object changes. (Setting 1)

> **Exploded** = Dimensions are not associated to the objects and are totally separate objects (arrows, lines and text). (Setting 0)

How to turn Associative Dimensioning On or OFF.

There are 2 methods:

Method 1.
On the command line type: ***dimassoc <enter>,***
then enter the number ***2, 1*** or ***0 <enter>.***

Method 2.
Select **Tools / Options / User Preferences tab.**

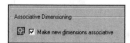

Checked box = On

Unchecked box = Off

Note: You cannot set Exploded here.

The Default setting is "2" or "ON", but check to make sure.

When you first enter AutoCAD, the True Associative dimensioning feature is "ON". This setting is saved with each individual drawing. It is not a system setting for all drawings. This means, when you open a drawing it is important to check the **dimassoc** command to verify that True Associative dimensioning is ON. Especially if you open a drawing created with an AutoCAD release previous to 2002.

How to Re-associate a dimension.

If a dimension was created with a previous version of AutoCAD or True Associative dimensioning was turned off, you may use the **dimreassociate** command to change the non-associative dimensions into associative dimensions.
(You may use the **Tools / Inquiry / List** command to determine whether a dimension is associative or non-associative.)

1. Select   **Dimension / Reassociate Dimensions**. (No icon available.)

   Command: _dimreassociate
   Select dimensions to reassociate ...

2. Select objects: *select the dimension to be reassociated.*

3. Select objects: *select more dimensions or <enter> to stop.*

4. Specify first extension line origin or [Select object] <next>: *an "X" will mark the first extension line; use object snap to select the exact location of the extension line point. (If the "X" has a box around it, just press <enter>.)*

5. Specify second extension line origin <next>: *the "X" will move to another location. Use object snap or <enter>.*

6. *Continue until all extension line points are selected.*

Note: You must use object snap to specify the exact location for the extension lines.

Regenerating Associative dimensions

Sometimes after panning and zooming, the associative dimensions seem to be floating or not following the object. The **DIMREGEN** command will move the associative dimensions back into their correct location.

*You must type **dimregen <enter>** on the command line. Sorry, AutoCAD did not provide an icon or a pull-down menu.*

LINEAR DIMENSIONING

First open the dimension toolbar. Using the toolbar icons to select the dimension commands is the most efficient method.

<u>Linear dimensioning</u> allows you to create horizontal and vertical dimensions.

1. Select the **LINEAR** command using the icon or Dimension / Linear.
   Command:__dimlinear
2. Specify first extension line origin or <select object>: *snap to first extension line origin (P1).*
3. Specify second extension line origin: *snap to second extension line origin (P2).*
4. Specify dimension line location or [Mtext/Text/Angle/Horizontal/Vertical/Rotated]: *select where you want the dimension line placed (P3).*
   Dimension text = 2.000 *(The dimension text value will be displayed on the last line.)*

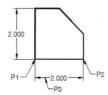

BASELINE DIMENSIONING

<u>Baseline dimensioning</u> allows you to establish a **baseline** for successive dimensions. The spacing between dimensions is automatic and should be set in dimension styles. A Baseline dimension must be used with an existing dimension. If you use Baseline dimensioning immediately after a Linear dimension, you do not have to specify the baseline origin.

1. Create a <u>linear</u> dimension first (1.400 P1 and P2).
2. Select the **BASELINE** command using the icon or Dimension / Baseline.
   Command: _dimbaseline
3. Specify a second extension line origin or [Undo/Select] <Select>: *snap to the second extension line origin (P3).*
   Dimension text = 2.588
4. Specify a second extension line origin or [Undo/Select] <Select>: *snap to P4.*
   Dimension text = 3.633
5. Specify a second extension line origin or [Undo/Select} <Select>: *select <enter> twice to stop.*

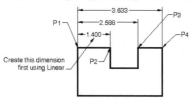

CONTINUE DIMENSIONING

Continue creates a series of dimensions in-line from an existing dimension. If you use the Continue dimensioning immediately after a Linear dimension, you do not have to specify the Continue extension origin.

Create a linear dimension first (1.400   P1 and P2).
Select the Continue command using the icon or Dimension / Continue.
Command: _dimcontinue
Specify a second extension line origin or [Undo/Select] <Select>: *snap to the second extension line origin (P3).*
Dimension text = 1.421
Specify a second extension line origin or [Undo/Select] <Select>: *snap to the second extension line origin (P4).*
Dimension text = 1.364
Specify a second extension line origin or [Undo/Select] <Select>: *press <enter> twice to stop.*

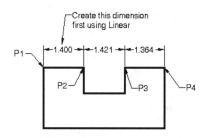

ALIGNED DIMENSIONING

The **ALIGNED** dimension command aligns the dimension with the angle of the object that you are dimensioning. The process is the same as Linear dimensioning. It requires two extension line origins and placement of text location. (Example below)

1. Select the **ALIGNED** command using one of the following:

   TYPE = DIMALIGNED or DIMALI or DAL
   PULLDOWN = DIMENSION / ALIGNED
   TOOLBAR = DIMENSION

   Command: _dimaligned
2. Specify first extension line origin or <select object>: *select the first extension line origin (P1)*
3. Specify second extension line origin: *select the second extension line origin (P2)*

4. Specify dimension line location or [Mtext/Text/Angle]: *place dimension text location*

   Dimension text = *the dimension value will appear here*

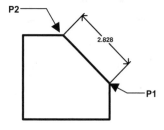

ANGULAR DIMENSIONING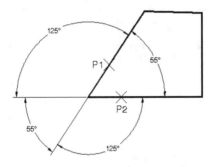

The **ANGULAR** dimension command is used to create an angular dimension between two lines that form an angle. All that is necessary is the selection of the two lines and the location for the dimension text.

The **degree symbol** is automatically added as the dimension is created.

1. Select the **ANGULAR** command.
   Command: _dimangular
2. Select arc, circle, line, or <specify vertex>: *click on the first line that forms the angle (P1) location is not important, do not use snap.*
3. Select second line: *click on the second line that forms the angle (P2)*
4. Specify dimension arc line location or [Mtext/Text/Angle]: *place dimension text location*
   Dimension text = *angle will be displayed here*

Any of the 4 angular dimensions shown below can be created by clicking on the 2 lines (P1 and P2) that form the angle.

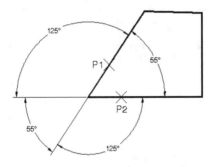

DIMENSIONING ARC LENGTHS – 2007 & 2006 only

You may dimension the distance along an Arc. This is known as the Arc length.
Arc length is an associative dimension.

Example:

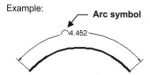

Arc symbol

4.452

Select the Arc length dimension command using one of the following:

TYPING = dimarc <enter>
PULLDOWNS = Dimension / Arc length
TOOLBARS =Dimension

Command: _dimarc
Select arc or polyline arc segment: *select the Arc*
Specify arc length dimension location, or [Mtext/Text/Angle/Partial/Leader]: *place the*
dimension line and text location
Dimension text = *dimension value will be shown here*

To differentiate the Arc length dimensions from Linear or Angular dimensions, arc length
dimensions display an arc ($\frown$) symbol by default. (Also called a "hat" or "cap")

The arc symbol may be displayed either <u>above,</u> or <u>preceding</u> the dimension text.
You may also choose not to display the arc symbol.

Example:

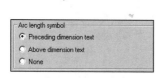

ABOVE   **PRECEDING**   **NONE**

Specify the placement of the arc symbol in the Dimension Style / Symbols and Arrows
tab or you may edit its position using the Properties Palette.

The extension lines of an Arc length dimension are displayed as <u>radial</u> if the included angle is <u>greater than 90 degrees</u>.

Example:

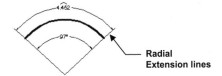

Radial
Extension lines

The extension lines of an Arc length dimension are displayed as <u>orthogonal</u> if the included angle is <u>less than 90 degrees</u>.

Example:

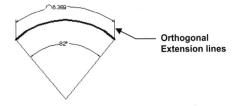

Orthogonal
Extension lines

DIMENSIONING A LARGE CURVE – 2007 & 2006 only

When dimensioning an arc the dimension line should pass through the center of the arc. However, for large curves, the center of the arc could be very far away. Even off the sheet.

When the true center location cannot be displayed you can create a "Jogged" radius dimension.

Example:

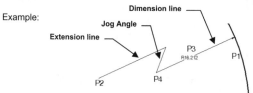

You can specify the jog angle in the Dimension Style / Symbols and Arrows tab.

1. Select the Jogged radius dimension using one of the following:

TYPING = dimjogged <enter>
PULLDOWNS = Dimension / jogged
TOOLBARS = Dimension

2. Select arc or circle: *select the large arc or circle (P1 anywhere on arc)*

3. Specify center location override: *move the cursor and left click to specify the new center location (P2).*

   Dimension text = (actual radius will be displayed here)

4. Specify dimension line location or [Mtext/Text/Angle]: *move the cursor and left click to specify the location for the dimension text (P3).*

5. Specify jog location: *move the cursor and left click to specify the location for the jog (P4).*

Options:

Mtext: Displays the In-Place Text Editor, which you can use to edit the dimension text.

Text: You may customize the dimension text on the command line. The actual dimension is displayed in brackets < >.

Angle: Changes the angle of the dimension text.

RADIAL DIMENSIONING

<u>DIAMETER</u> DIMENSIONING

The **DIAMETER** dimensioning command should be used when dimensioning circles and arcs of <u>more than 180 degrees</u>.

<u>Center marks</u> are automatically drawn as you use the diameter dimensioning command. If the circle already has a center mark or you do not want a center mark, set the center mark setting to **<u>NONE</u>** (Dimension Style / Symbols and Arrows tab) before using Diameter dimensioning.

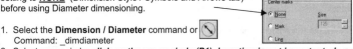

1. Select the **Dimension / Diameter** command or
   Command: _dimdiameter
2. Select arc or circle: *click on the arc or circle (P1). Location is not important; <u>do not use object snap.</u>*
   Dimension text = *the diameter will be displayed here.*
3. Specify dimension line location or [Mtext/Text/Angle]: *place dimension text location (P2).*

P2 — Ø2.000

P1

Note: Version 2006 and 2007 users
You may **FLIP** the direction of the arrowhead using the following method:
1. Select the dimension
2. Press the right mouse button.
3. Select "**Flip Arrow**" from the menu.

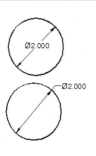

Ø2.000

Ø2.000

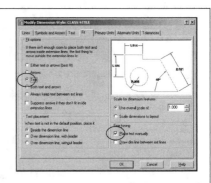

f you would like your Diameter dimensions to appear as shown in the two examples above, you must change the "**<u>Fit Options</u>**" and "**<u>Fine Tuning</u>**" in the **<u>Fit tab</u>** in your Dimension Style.

RADIAL DIMENSIONING (continued)

RADIUS DIMENSIONING

The **RADIUS** dimensioning command should be used when dimensioning arcs of <u>LESS than 180 degrees.</u>

<u>Center Marks</u> are automatically drawn as you use the **RADIUS** dimensioning command. If the circle already has a center mark, set the center mark to **NONE** (in the Dimension Style) before using RADIUS dimensioning. Or create a "sub-style".

1. Select the **Dimension / Radius** command or ⊙
   Command: _dimradius
2. Select arc or circle: **click on the arc (P1). Location is not
   important, do not use snap.**
   Dimension text = **the radius will be displayed here**
3. Specify dimension line location or [Mtext/Text/Angle]: **place dim text location (P2)**

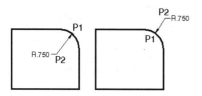

> **Note: Version 2006 and 2007 users**
> You may **FLIP** the direction of the
> arrowhead using the following method:
> 1. Select the dimension
> 2. Press the right mouse button.
> 3. Select "**Flip Arrow**" from the menu.

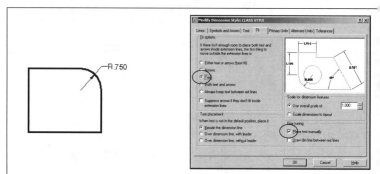

If you would like your Radius dimension to appear as shown in the example immediately above, you must change the "**Fit Options**" and "**Fine Tuning**" in the **Fit tab** in your Dimension Style.

QUICK DIMENSION

Quick Dimension creates multiple dimensions with one command. Quick Dimension can create Continuous, Staggered, Baseline, Ordinate, Radius and Diameter dimensions. Qdim only works in modelspace. It is not trans-spatial.

1. Select the **Quick Dimension** command using one of the following:

> **TYPE = QDIM**
> **PULLDOWN = Dimension/Quick Dimension**
> **TOOLBAR = DIMENSION**

> Command: _qdim
> Associative dimension priority = Endpoint
> Select geometry to dimension

CONTINUOUS

2. Select the objects to be dimensioned with a crossing window or pick each object
3. Press <enter> to stop
4. Specify dimension line position, or
[Continuous/Staggered/Baseline/Ordinate/Radius/Diameter/datumPoint/Edit/settings]
<Continuous>: *Select "C" <enter> for Continuous.*
5. Select the location of the dimension line.

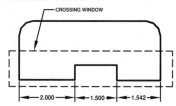

BASELINE

2. Select the objects to be dimensioned with a crossing window or pick each object.
3. Press <enter> to stop
4. Specify dimension line position, or
[Continuous/Staggered/Baseline/Ordinate/Radius/Diameter/datumPoint/Edit/settings]
<Continuous>: *Select "B" <enter> for Baseline.*
5. Select the location of the dimension line.

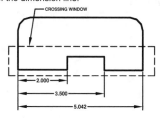

STAGGERED

2. Select the objects to be dimensioned with a crossing window or pick each object.
3. Press <enter> to stop
4. Specify dimension line position, or
   [Continuous/Staggered/Baseline/Ordinate/Radius/Diameter/datumPoint/Edit/settings]
   <Continuous>: *Select "S" <enter> for Staggered.*
5. Select the location of the dimension line.

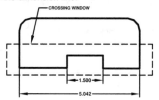

DIAMETER

2. Select the objects to be dimensioned with a crossing window or pick each object
   (Qdim will automatically filter out any linear dims)
3. Press <enter> to stop.
4. Specify dimension line position, or
   [Continuous/Staggered/Baseline/Ordinate/Radius/Diameter/datumPoint/Edit/settings]
   <Continuous>: *Select "D" <enter> for Diameter*
5. Select the location of the dimension line.
(Dimension line length is determined by the "Baseline Spacing" setting)

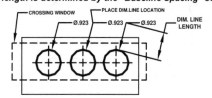

RADIUS

2. Select the objects to be dimensioned with a crossing window or pick each object.
   (Quick Dimension will automatically filter out any linear dimensions)
3. Press <enter> to stop selecting objects
4. Specify dimension line position, or
   [Continuous/Staggered/Baseline/Ordinate/Radius/Diameter/datumPoint/Edit/settings]
   <Continuous>: *Select "R" <enter> for Radius*
*The dimensions are automatically placed, you do not select the location for the
dimension line. Dimension line length is determined by the "Baseline Spacing" setting)*

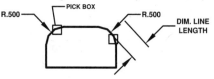

LEADER

The **LEADER** command is primarily used to add a **note** to an object. A Leader's appearance is very similar to a Radial dimension **but** a Leader **should not** to be used for Radial dimensioning. A Leader allows you to type a note at the end of the hook. (Note: Leader and Qleader are the same command)

If associative dimensioning is turned on with DIMASSOC, the leader start point can be associated with a location on an object. If the object is relocated, the arrowhead remains attached to the object and the leader line stretches, but the text or feature control frame remains in place.

1. Select the **LEADER** command using one of the following.

> **TYPE = LE or QLEADER**
> **PULLDOWN = DIMENSION / LEADER**
> **TOOLBAR = DIMENSION**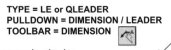

See "Settings" below

Command: _qleader
2. Specify first leader point, or [Settings]<Settings>: *select the location for the arrowhead (P1)*
3. Specify next point: *select next location (P2)*
4. Specify next point: *select another point or press <enter>*
5. Specify text <u>width</u> <.000>: *press <enter> (NOTE: This is not asking for Height)*
6. Enter first line of annotation text <Mtext>: *type your desired text here*
7. Enter next line of annotation text: *type more text or press <enter> to stop*

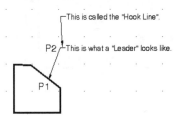

This is called the "Hook Line".

P2 — This is what a "Leader" looks like.

P1

HOOK LINE
The hook line is automatically added to the last line segment (P2) if the leader line is 15 degrees or more from horizontal. The length of the Hook Line is controlled by the arrow length setting in the dimension style.

SETTINGS:
If you would like to make changes to the appearance of the Leader, select **"Settings"** before you place the first point (arrowhead location). Selecting **"Settings"** will display a dialog box with many options.

LINE WITH ARROW ONLY (NO TEXT)

Occasionally you will need to draw a line with an arrow on the end. But you do not want text at the other end. (Just a line with an arrow pointing at something within your drawing.)

To accomplish this you must tell AutoCAD that you do not want text. (AutoCAD calls this "annotation") You must change the "annotation type " setting to "None".

CHANGING THE SETTING FOR ANNOTATION TYPE

1. Select the **LEADER** command using one of the following.

> **TYPE = LE or QLEADER**
> **PULLDOWN = DIMENSION / LEADER**
> **TOOLBAR = DIMENSION**

Command: _qleader
2. Specify first leader point, or [Settings]<Settings>: *select the "Settings" option*

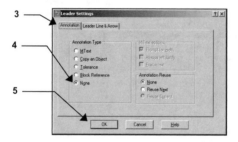

3. Select the "**Annotation**" tab.
4. Select "**None**"
5. Select the "**OK**" button.

Now continue drawing the arrow and line.

6. Specify first leader point, or [Settings]<Settings>: *select the location for the arrowhead (P1)*
7. Specify next point: *select the location for the other end of the line (P2)*
8. Specify next point: *press <enter> to stop*

P1 ◀━━━━━━━━━━━━━━▶ P2

Note: You will have to change Annotation Type back to "Mtext" if you wish to create a leader with text.

RDINATE Dimensioning

nate dimensioning is primarily used by the sheet metal industry. But many others are izing the speed and tidiness this dimensioning process allows.

inate dimensioning is used when the X and the Y coordinates, from one location, are the dimensions necessary. Usually the part has a uniform thickness, such as a flat plate with s drilled into it. The dimensions to each feature, such as a hole, originate from one um" location. This is similar to "baseline" dimensioning. Ordinate dimensions have only datum. The datum location is usually the lower left corner of the object.

inate dimensions appearance is also different. Each **dimension** has only one **leader line** a **numerical value**. Ordinate dimensions do not have extension lines or arrows.

mple of Ordinate dimensioning:

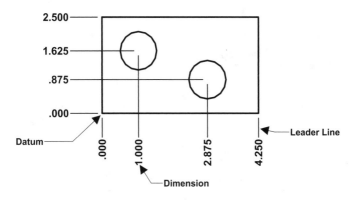

e:
inate dimensions can be Associative and are Trans-spatial. Which means that you dimension in paperspace and the ordinate dimensions will remain associated to the ect they dimension. (Except for Qdim ordinate)

er to the next page for step by step instructions to create Ordinate dimensions.

Creating Ordinate dimensions

1. Move the "Origin" to the desired "datum" location as follows:
   Note: This must be done in Model Space.
   a. Select **TOOLS / MOVE UCS**
   b. Snap to the desired location.

2. Select the Ordinate command using one of the following:
   Note: Ordinate dimensioning may be used in model or paper space.

   TYPE = DIMORDINATE
   PULLDOWN = DIMENSION / ORDINATE
   TOOLBAR = DIMENSION

3. Select the first feature, using object snap.

4. Drag the leader line horizontally or vertically away from the feature.

5. Select the location of the "leader endpoint.
   (The dimension text will align with the leader line)

Use "Ortho" to keep the leader lines straight.

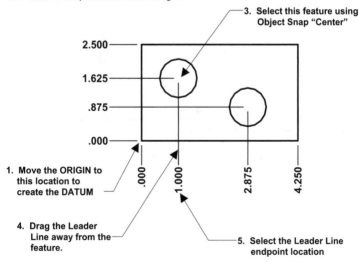

3. **Select this feature using Object Snap "Center"**

1. **Move the ORIGIN to this location to create the DATUM**

4. **Drag the Leader Line away from the feature.**

5. **Select the Leader Line endpoint location**

ere is insufficient room for a dimension you may want to jog the dimension.
"jog" the dimension, as shown below, turn "**Ortho**" **off** before placing the Leader Line
point location. The leader line will automatically jog. With Ortho off, you can only indicate
feature location and the leader line endpoint location, the leader line will jog the way it
its to.

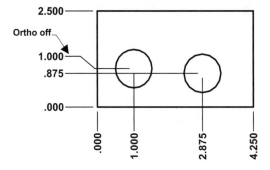

Qdim with Ordinate dimensioning (Not available in version LT)

1. Select **DIMENSION / QDIM**
2. Select the geometry to dimension <enter>
3. Type **"O"** <enter> to select Ordinate
4. Type **"P"** <enter> to select the **datumPoint** option.
5. Select the datum location on the object. (use Object snap)
6. Drag the dimensions to the desired distance away from the object.

> **Note: Qdim can be associative but is not trans-spatial.**
> **If the object is in Model Space, you must dimension in Model Space.**

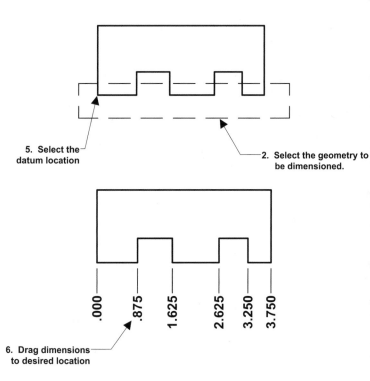

5. Select the datum location

2. Select the geometry to be dimensioned.

6. Drag dimensions to desired location

)LERANCES

en you design and dimension a widget, it would be nice if when that widget was made, all the dimensions were exactly as you had asked. But in reality this is very difficult and or ensive. So you have to decide what actual dimensions you could live with. Could the get be just a little bit bigger or smaller and still work? This is why tolerances are used.

olerance is a way to communicate, to the person making the widget, how much larger or aller this widget can be and still be acceptable. In other words each dimension can be en a maximum and minimum size. But the widget must stay within that **"tolerance"** to be rect. For example: a hole that is dimensioned 1.00 +.06 -.00 means the hole is nominally 0 but it can be as large as 1.06 but can not be smaller than 1.00.

Select **DIMENSION / STYLE / MODIFY**
Select the **TOLERANCES UNITS** tab.

> *Note: if the dimensions in the display look strange,*
> *make sure "Alternate Units" are turned OFF.*

Method
e options allows you to select how you would like the tolerances displayed. There are 5 thods: None, Symmetrical, Deviation, Limits. (Basic is used in geometric tolerancing and not be discussed at this time)

fer to the next page for descriptions of methods. —— 3

Scaling for height. This ntrols the height of the erance text. The entered ue is a percentage of the mary text height. If .50 is tered, the tolerance text ght will be 50% of the mary text height.

— 4

— 5

Vertical position. This ntrols the placement of the erance text in relation to the mary text. The options are p, Middle and Bottom. hichever option you select, it l align the tolerance text with e bottom of the primary text.

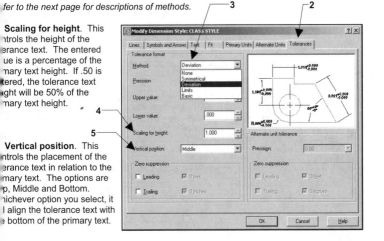

SYMMETRICAL is an equal bilateral tolerance. It can vary as much in the plus as in the negative. Because it is equal in the plus and minus direction, only the "Upper value" box is used. The "Lower value" box is grayed out.

Example of a Symmetrical tolerance
:

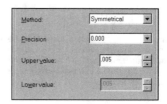

DEVIATION is an unequal bilateral tolerance. The variation in size can be different in both the plus and minus directions. Because it is different in the plus and the minus the "Upper" and "Lower" value boxes can be used.

Example of a Deviation tolerance:

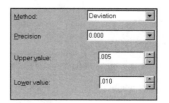

Note: If you set the upper and lower values the same, the tolerance will be displayed as symmetrical.

LIMITS is the same as deviation except in how the tolerance is displayed. Limits calculates the plus and minus by adding and subtracting the tolerances from the nominal dimension and displays the results. Some companies prefer this method because no math is necessary when making the widget. Both "Upper and Lower" value boxes can be used.
Note: The "Scaling for height" should be set to "1".

Example of a Limits tolerance:

GEOMETRIC TOLERANCING

ometric tolerancing is a general term that refers to tolerances used to control the form,
•file, orientation, runout, and location of features on an object. Geometric tolerancing is
marily used for mechanical design and manufacturing. The instructions below will cover the
•erance command for creating geometric tolerancing symbols and feature control frames.

you are not familiar with geometric tolerancing, you may choose to skip this lesson.

Select the **TOLERANCE** command using one of the following:

TYPE = TOL
PULLDOWN = DIMENSION / TOLERANCE
TOOLBAR = DIMENSION

The Geometric Tolerance dialog box, shown below, should appear.

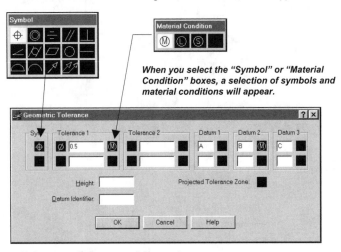

*When you select the "Symbol" or "Material
Condition" boxes, a selection of symbols and
material conditions will appear.*

Make your selections and fill in the tolerance and datum boxes.
Select the **OK** box.
The tolerance should appear attached to your cursor. Move the cursor to the desired
location and press the left mouse button.

•te: *the size of the Feature Control Frame above, is determined by the height of the
mension text.*

3-23

GEOMETRIC TOLERANCES and QLEADER

The **Qleader** command allows you to draw leader lines and access the dialog boxes used to create feature control frames in one operation.

1. Select **Dimension / Leader**
2. Select the **"Settings"** option. (Right click, and select Settings from the short cut menu)

The Leader Settings dialog box should appear.

3. Select the Annotation tab.
4. Select Tolerance then OK.

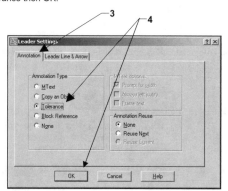

5. Place the first leader point. **P1**
6. Place the next point **P2**
7. Press <enter>

The Geometric Tolerance dialog box will appear.

8. Make your selections and fill in the tolerance and datum boxes.

9. Select the **OK** button.

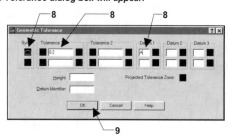

ATUM FEATURE SYMBOL

atum in a drawing is identified by a **"datum feature symbol".**

create a *datum feature symbol*:

Select Dimension / Tolerance
Type the "datum reference letter" in the "Datum Identifier" box.
Select the OK button.

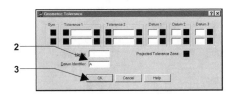

create a *datum feature symbol* combined with a *feature control frame*:

Select **Dimension / Tolerance**
Make your selections and fill in the tolerance.
Type the "datum reference letter" in the "Datum Identifier" box.
Select the **OK** button.

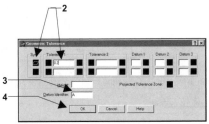

DATUM TRIANGLE

A datum feature symbol, in accordance with ASME Y14.5M-1994, includes a leader line and a datum triangle filled. You can create a wblock or you can use the two step method below using Dimension / Tolerance and Qleader.

1. Select **Dimension / Leader**.
2. Select the **"Settings"** option. (Right click, and select Settings from the short cut menu)

The Leader Settings dialog box should appear.

3. Select the Annotation tab.
4. Select the Tolerance button.

5. Select the Leader Line & Arrow tab.
6. Select Datum Triangle Filled
7. Select the OK button.

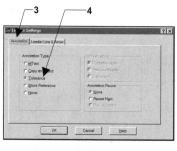

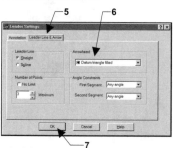

8. Specify the first leader point. (The triangle endpoint)
9. Specify next point then press <enter>
10. When the Geometric Tolerance dialog box appears, select the OK button.

*If you were successful,
a <u>datum triangle filled</u> with
a <u>leader line</u> should appear.
(As shown below)*

Note: When you change the leader settings AutoCAD automatically creates an "Override" in dimensions and "sets it current". To make changes to the Leader settings you must change the "override".

11. Next create a datum feature symbol. (Follow the instructions on the previous page.)

12. Now move the datum feature symbol to the endpoint of the leader line to create the symbol below left.

You are probably wondering why we didn't just type "A" in the identifier box. That method will work if your leader line is horizontal. But if the leader line is vertical, as shown on the left, it will not work. (The example on the right illustrates how it would appear)

'PING GEOMETRIC SYMBOLS

)u want geometric symbols in the notes that you place on the drawing, you can easily
omplish this using a font named **GDT.SHX**. This font will allow you to type normal letters
geometric symbols, in the same sentence, by merely toggling the SHIFT key up and down
CAPS LOCK on.

First you must create a new text style using the **GDT.SHX** font.

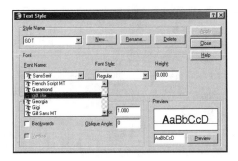

CAPS LOCK must be **ON**.

Select **DRAW / TEXT / Single Line or Multiline**

Now type the the sentence shown below. When you want to type a symbol, press the
SHIFT key and type the letter that corresponds to the symbol. For example: If you want the
diameter symbol, press the shift key and the "N" key. (Refer to the alphabet of letters and
symbols shown above.)

3X ⌀.44   ⊔⌀1.06 ▽.06

Can you decipher what it says?
(Drill (3) .44 diameter holes with a 1.06 counterbore diameter .06 deep)

MODIFY AN <u>ENTIRE</u> DIMENSION STYLE

After you have created a Dimension Style, you may find that you have changed your mind about some of the settings. You can easily change the entire Style by using the "Modify" button in the Dimension style Manager dialog box. This will not only change the Style for future use, but it will also <u>update dimensions already in the drawing</u>.

Note: if you do not want to update the dimensions already in the drawing, but want to make a change to the next dimension drawn, refer to Override.

1. Select the Dimension Style command.

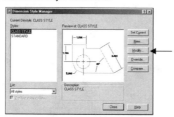

2. Select the Modify button from the Dimension Style Manager dialog box.

3. Make the desired changes to the settings.

4. Select the OK button.

5. Select the Close button.

Now look at your drawing. Have your dimensions updated?

Note: If some of the dimensions have not changed:
1. Select   Dimension / Update
2. Select the dimension(s) you wish to update and press <enter>

Note: This feature <u>will not</u> change dimensions that have been modified or exploded.

EDITING A DIMENSION using PROPERTIES PALETTE

Sometimes you would like to modify the settings of an **individual existing** dimension. This can be achieved using the Modify Properties command.

1. Double click on the dimension that you wish to change.

   The Properties Palette will appear.

2. Select and change the desired settings.

   Example: Change the dimension text height or size of the Arrow.

3. Press <enter> . *(The change should have taken effect)*

4. Press the <esc> key

Note: **The dimension will remain Associative.**

Note: **This process <u>will not</u> change <u>Exploded</u> dimensions.**

EDITING DIMENSION TEXT VALUES — Version 2005 & 2004

Sometimes you need to modify the dimension text. You may add a symbol, a note or even change the text of an existing dimension. There are 2 methods.

Example: Add the word "Max." to the existing dimension.

Before Editing   **After Editing**

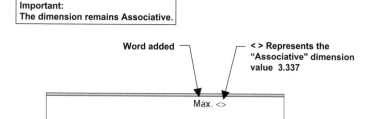

Method 1.
1. Double click on the dimension that you want to change.
   The Properties Palette will appear.
2. Scroll down to Text / Text override ────────
   (Notice that the actual measurement is directly above it.)
3. Type the new text and press <enter>

Important:
If you use this method the dimension
is no longer Associative.

Method 2.
1. Type on the Command line: **DDEDIT** <enter>
2. Select the dimension you want to edit. (Only one at a time)
3. The "Multiline Text Editor" will appear.
   If the dimension is <u>Associative</u>, it will display the dimension text as a symbol <>.
   Do not erase this symbol. You can add text before or after the symbol.
4. Select the OK button.

Important:
The dimension remains Associative.

Word added ─┐   ┌─ **< > Represents the**
                                    "Associative" dimension
                                    value 3.337

Max. <>

EDITING DIMENSION TEXT VALUES – Version 2006

Sometimes you need to modify the dimension value text. You may add a symbol, a note or even change the text of an existing dimension. The following describes 2 methods.

Example: Add the word "Max." to the existing dimension value text.

Before Editing

-3.337-

After Editing

-Max. 3.337-

Method 1 (Properties Palette).
1. Double click on the dimension that you want to change.
   The Properties Palette will appear.
2. Scroll down to **Text / Text override**
   (Notice that the actual <u>measurement</u> is directly above it.)
3. Type the new text (Max) and **< >** and press <enter>
   (< > represents the associative text value)

**Important:
If you do not use < > the dimension will no longer be Associative.**

Method 2 (ddedit).
1. Type on the Command line: **ed** <enter> (This is the **ddedit** command)
2. Select the dimension you want to edit.
3. The "In-Place Text Editor" will appear.

   Associative Dimension
   If the dimension is <u>Associative</u> the dimension text will appear highlighted.

   You may add text in front or behind the dimension text and it will remain Associative.
   Be careful not to disturb the dimension value text.

   -Max 9.815-

   Non Associative Dimension
   If the dimension value has been changed or exploded it will appear with a gray background and is not Associative.

   -9.815 Max-

4. Make the change.
5. Select the OK button.

EDITING THE DIMENSION POSITION

Sometimes dimensions are too close and you would like to stagger the text or you need to move an entire dimension to a new location, such as the examples below.

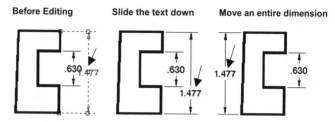

Before Editing   **Slide the text down**   **Move an entire dimension**

Editing the position is easy.
1. Select the dimension that you want to change. (Grips will appear)
2. Select the middle grip. (It will become red and active, hot)
3. Move the cursor and the dimension will respond.
4. Press the left mouse button to place the dimension in the new location.

ADDITIONAL EDITING OPTIONS USING THE SHORTCUT MENU

1. Select the dimension that you want to change.
2. Press the Right Mouse button.
3. Select "Dimension Text Position" from the shortcut menu shown to the right.
4. A sub-menu appears with many editing options.

Experiment with these options. They will be very helpful in the future.

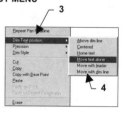

Note: This command will not work with <u>Exploded</u> dimensions.
You will have to use the "<u>Stretch</u>" command to achieve the same results.

EDITING MULTIPLE DIMENSIONS

You can edit existing multiple dimensions using the QDIM / Edit command. The Qdim, edit command will edit all multiple dimensions, no matter whether they were created originally with Qdim or not. All multiple: linear, baseline and continue dimensions respond to this editing command.

1. Select the **QDIM** command.
2. Select the dimensions to edit.

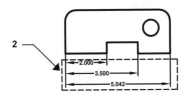

3. Press <enter> to stop selecting.
4. Select "E" for edit.

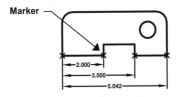

Small *markers* will appear at the extension line origins.
These markers are called **"dimension points"**.
You can **Add** or **Remove** dimensions by selecting these markers.

TO <u>REMOVE</u> A DIMENSION

AutoCAD assumes that you want to Remove dimensions, so this is the default setting.

5. Indicate dimension point to remove, or [Add/eXit] <eXit>: **Click on the extension line markers you want to remove**

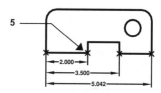

6. Press "X" to stop selecting.
7. Move the cursor to place the remaining dimensions again.

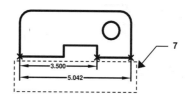

TO <u>ADD</u> A DIMENSION

8. Select "A" <enter> for Add
9. Using Object Snap, select the extension line origins of the dimension you want to Add.
10. Press "X" to stop selecting.
11. Replace the dimension line location.

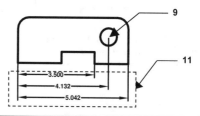

Why would I bother using ADD and REMOVE when I can just use Erase or add a dimension? The Qdim ADD and REMOVE options automatically re-space the dimensions. You will not have to stretch or move the remaining dimensions.

DIMENSION STYLES

Using the "Dimension Style Manager" you can change the appearance of the dimension features, such as length of arrowheads, size of the dimension text, etc. There are over 70 different settings.
You can also Create New, Modify, Override and Compare Dimension Styles. All of these are simple, by using the Dimension Style Manager described below.

Select the "Dimension Style Manager" using one of the following:

TYPE = DDIM
PULLDOWN = DIMENSION / STYLE
TOOLBAR = DIMENSION

The following dialog box will appear.

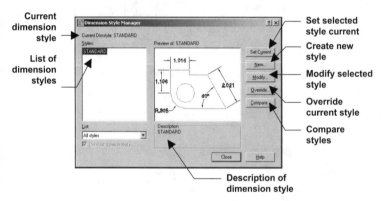

Current dimension style

List of dimension styles

Set selected style current

Create new style

Modify selected style

Override current style

Compare styles

Description of dimension style

Set Current Select a style from the list of styles and select the **set current** button.
(Only Standard is shown unless you have previously created other styles.)

New Select this button to create a new style. When you select this button, the **Create New Dimension Style** dialog box is displayed.

Modify Selecting this button opens the **Modify dimension Style** dialog box which allows you to make changes to the current style.

Override An override is a temporary change to the current style. Selecting this button opens the Override Current Style dialog box.

Compare Compares two styles.

CREATING A NEW DIMENSION STYLE

When creating a new style you must start with an existing style, such as Standard.
Next, assign it a new name, make the desired changes and when you select the OK
button the new style will have been successfully created.

LET'S CREATE A NEW DIMENSION STYLE.
A dimension style is a group of settings that has been saved with a name you assign.

1. Open your drawing **BSIZE** and set **DIMASSOC** to **2.**
2. Select the **DIMENSION STYLE** command
3. Select the **NEW** button
4. Enter **CLASS STYLE** in the "New Style Name" box.
5. **"Start With:" box**: Start with the settings in the STANDARD style and then you will make some changes.
6. **"Use For:" box** is for creating "family" dimension styles and will be discussed later. For now, leave it set to "All dimensions".
7. Select the **CONTINUE** button.

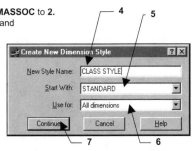

The "New Dimension Style: Class Style" dialog box will appear.

8. Select the **"Primary Units"** tab and make the changes shown below.

NOTE: REFER TO "DIMENSION STYLE DEFINITIONS" for descriptions of settings.

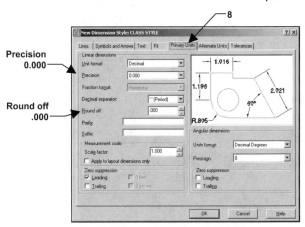

DO NOT SELECT THE OK BUTTON YET.

9. Select the **"Lines"** tab and make the changes shown below.

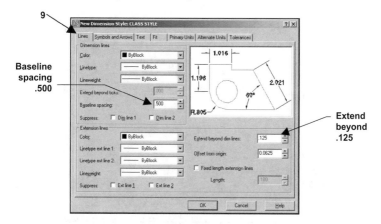

DO NOT SELECT THE OK BUTTON YET.

10. Select the **"Symbols and Arrows"** tab and make the changes shown below.

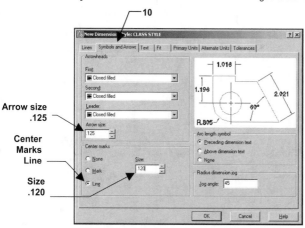

DO NOT SELECT THE OK BUTTON YET.

11. Select the **"Text"** tab.
— 11

Text Height
.125

Offset
.060

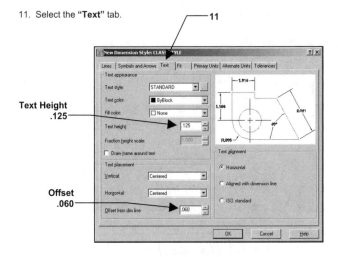

DO NOT SELECT THE OK BUTTON YET.

12. Select the **"Fit"** tab.
— 12

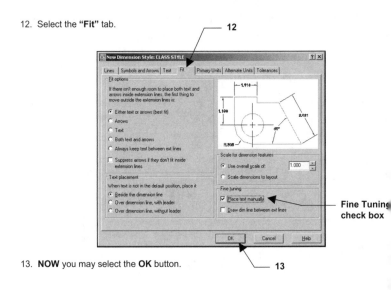

Fine Tuning
check box

13. **NOW** you may select the **OK** button.
— 13

3-38

14. *Your new style "Class Style" should be listed.* Select the **"Set Current"** button to make your new style "Class Style" the style that will be used.

Your new — style is listed here.

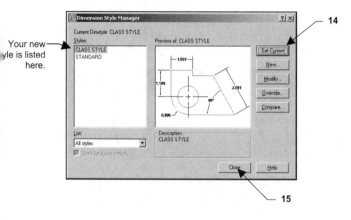

14

15

15. Select the **Close** button to **exit**.

16. Save your drawing

Note: You have successfully created a new "Dimension Style" called "Class Style". This style will be saved in the current drawing after you save the drawing. It is important that you understand that this dimension style resides only in the current drawing. If you open another drawing, this dimension style will not be there.

COMPARE TWO DIMENSION STYLES

Sometimes it is useful to compare the settings of two styles. Compare will compare the two styles and list the differences.

1. Select the Dimension Style command.
2. Select the **Compare** button.

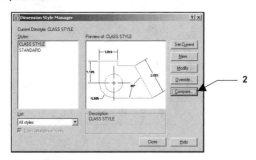

The "Compare Dimension Style" dialog box will appear.

3. Select a style such as **Class Style** in the **Compare** box.
4. Select **Standard** in the **With** box.
5. AutoCAD found differences and listed them.
6. Select "Close" button to exit.

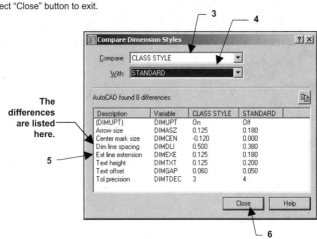

CREATING A DIMENSION SUB-STYLE

Sometimes when using a dimension style, you would like the Linear, Angular, Diameter and Radius dimensions to have different settings. But you want them to use the same dimension style. To achieve this, you must create a "sub-style".
Sub-styles have also been called "children" of the "Parent" dimension style. As a result, they form a family.
A Sub-style is permanent, unlike the Override command, which is temporary.

LET'S CREATE A SUB-STYLE FOR RADIUS.

We will set the center mark to None for the Radius command only.
The Diameter command center mark will not change.

1. Open your **BSIZE** drawing.

2. Select the **DIMENSION / STYLE** command.

3. Select **"Class Style"** from the Style List.

4. Select the **NEW** button.

5. Change the "Use for" to:
   Radius Dimensions

6. Select **Continue**

Version 2007 & 2006
7. Select the **Symbols and Arrows** tab.

Version 2005 & 2004
7. Select the **Lines and Arrows** tab.

8. Change the "Center Marks for Circles" Type: **NONE**

9. Select the **OK** button

This will turn gray. That's OK.

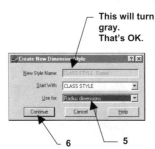

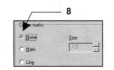

You have now created a sub-style that will automatically override the basic "Class Style" whenever you use the Radius command.

Current Dimstyle: CLASS STYLE
Styles:
CLASS STYLE
▶ Radial
STANDARD

OVERRIDE A DIMENSION STYLE

A dimension Override is a **temporary** change to the dimension settings.
An override **will not affect existing dimensions**. It will affect **new dimensions only**.

*Use this option when you want your **next dimension** just a little bit different but you don't
want to create a whole new dimension style and you don't want the existing dimensions to
change either.*

1. Select **Format / Dimension Style.**

2. Select the "**Style"** you want to override. (such as Class Style)

3. Select the **Override** button.

4. Make the desired changes to the settings.

5. Select the OK button.

6. Confirm the Override
   a. Look at the List of styles. Under the Style name, a sub heading of "Style overrides"
      should be displayed.
   b. The description box should display the style name and the override settings.

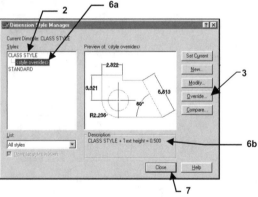

7. Select the Close button.

If you want to return to the style such as "**Class Style"**, select "Class Style" and then
select the "**Set Current"** button. Each time you select a different style, you must select
the **Set Current** button to activate it.

ALTERNATE UNITS

The options in this tab allow you to display inches as the primary units and the millimeter equivalent as alternate units. The millimeter value will be displayed inside brackets immediately following the inch dimension. Example: 1.00 [25.40]

Select **DIMENSION / STYLE / MODIFY**
Select the **ALTERNATE UNITS** tab.

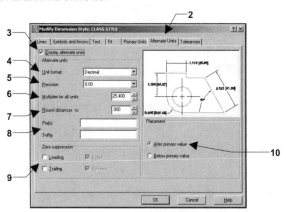

Display alternate units. Check this box to turn ON alternate units

Unit format. Select the Units for the alternate units.

Precision Select the Precision of the alternate units. This is independent of the Primary Units.

Multiplier for all units The primary units will be multiplied by this number to display the alternate unit value.

Round distance to Enter the desired increment to round off the alternate units value.

Prefix / Suffix This allows you to include a Prefix or Suffix to the alternate units. Such as: type **mm** to the Suffix box to display **mm** (for millimeters) after the alternate units.

Zero Suppression If you check one or both of these boxes, it means that the zero will not be drawn. It will be suppressed.

Placement Select the desired placement of the alternate units. Do you want them to follow immediately after the Primary units or do you want the Alternate units to be below the primary units?

Note: You alternate units are turned off you may display them by editing the dimension text in properties.
Use <> to represent the associative dimension and [ ] to display the alternative dimension.
Example: If the dimension is **15.00** you would enter **< > [ ]**.
The display would change to **15.00 [381]**.
Must be used with the generated dimension text. Cannot be displayed alone.

Dimension Style Definitions
Version 2007 & 2006 (2005 and 2004 are very similar but not shown)

When you select **Dimension / Style / Modify** the following dialog box will appear.

The following are descriptions for each setting within each section tab.

Lines tab

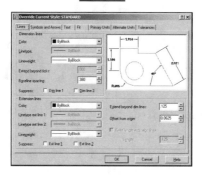

Dimension Lines
Color: Color of the dimension line.
Linetype: Sets the linetype for the dimension lines.
Lineweight: Sets the width of the dimension lines.
Extend beyond ticks: Distance to extend <u>dimension</u> line beyond the <u>extension</u> line.
Baseline spacing: Spacing between baseline dimensions.
Suppress: Suppress means disappear. You may suppress individually or both.

Extension Lines
Color: Color of the extension lines
Linetype ext line 1: Sets the linetype for the 1st extension lines.
Linetype ext line 2: Sets the linetype for the 2nd extension lines.
Lineweight: Sets the width of the extension lines.
Supress: You may suppress the 1st or 2nd extension lines individually or both.
Extend beyond dim. line: Distance to extend <u>extension</u> line beyond the <u>dimension</u> line. Example above.
Offset from origin: The distance between the object and the extension line. Example above.
Fixed length extension lines: Sets the total length of the extension lines starting from the dimension line toward the dimension origin.

Symbols and Arrows tab

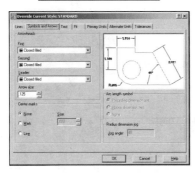

Arrowheads
First : Sets the style for the symbol inserted for the first extension point
Second: Sets the style for the symbol inserted for the second extension point and for
Diameter and Radius.
Leader: Sets the style for the symbol inserted for Leaders.
Arrow Size: Sets the Size of Arrowheads.

Center Marks for Circles
None: No center marks.
Mark: criss cross in the center of the circle
Line: Mark plus lines extending beyond the circle.
Size: Specify Size.

Text Appearance

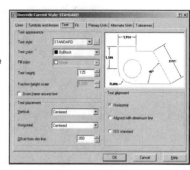

Text Style: Select which text style to use
for the dimension text. To create or
change a style, select the [...] button.
Text Color: Sets the color of the
dimension text.
Fill Color: Sets the color for the text
background in dimensions.
Text height: Sets the height.
Fraction Height Scale: Sets the size of
fractions relative to the dimension text
height. This is a factor not and actual
height. Example: A setting of .50 would be half of the dimension text height.
Draw Frame Around Text: If this box is checked, a box will be drawn around the text.

Text Placement

Vertical
Centered: Centers dimension text between extension lines.
Above: Places dimension text above the dimension line.
Outside: Places dimension text on the side of the dimension line farthest away from
the object.

Horizontal
Centered: Centers the dimension text along the dimension line between the ext
lines.
At Ext Line 1: Moves text near first extension line.
At Ext Line 2: Moves text near second extension line.
Over Ext Line 1: Places text over first extension line.
Over Ext Line 2: Places text over second extension Line

Offset from Dim Line: Sets the gap between the dimension text and the dimension
line.

Text Alignment
Horzontal: Places dimension text horizontal.
Aligned: Aligns dimension text with the dimension line.
ISO Standard: Aligns text with the dimension line when text is inside the extension
lines, but aligns it horizontally when text is outside extension lines.

FIT tab

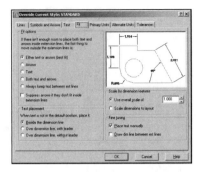

Fit Options: Controls the placement of text and arrowheads based on the space available between the extension lines. If there is not enough room to place both text and arrows inside extension lines, you must choose what to move outside the extension lines.

Text Placement: When dimension text is moved from its dim. style location, this setting controls the placement.

Scale for Dimension Features:

Use overall scale of: Set the factor for all dimension settings. Example: if this is set to 10 and the dimension text height is set to 1/8, the dimension text height would appear 1-1/4" ht. (10 x 1/8")

Scale dimensions to layout (paper space): Calculates a scale factor based on the scaling of model space vs. paper space.

Fine Tuning

Place text manually when dimensioning: You control the placement of the text if this feature is ON. It is not automatic.

Draw dim Line between Ext Lines: Draws the dimension line inside the extension lines even if the arrows are on the outside.

Linear Dimensions

Unit Format: Sets the units format for all dimensions except Angular.
Precision: Sets the number of decimal places in the dimension.
Fraction Format: Set the format for fraction to Horizontal, Diagonal or not stacked. Only available if Unit Format (above) is set to "Fractional".
Decimal Separator: Sets the style for the decimal separator to period, comma or space. Not available if Unit Format is set to Fractional.

Round Off: Sets rounding limits for dimensioning, such as .000 or 1/8. Rounds **up** to the nearest 3 place decimal or nearest 1/8".
Prefix: Add text in front of the dimension text. (**Flat for** 2.00)
Suffix: Add text after the dimension text. (1'-0" **Max**)

Measurement scale:

Scale Factor: AutoCAD multiples the dimension measurement by the value entered here. Example: If you draw a 1/2 inch line. Set this feature to 2. When you dimension the line, the dimension text will display 1. (2 X 1/2) If you set this feature to .50, when you dimension the line, the dimension text will display 1/4. (.50 X 1/2)
Apply to Layout Dimensions Only: Unnecessary now that we have Trans-spatial dimensioning.

Zero Suppression

The following two only work with decimals:
Leading: Controls the display of zeros before the decimal point. 0.50 = Off   .50 = ON
Trailing: Controls the display of zeros at the end of the dimension. .500 = Off .5 = ON

The following two only work with architectural:
Feet: Controls the display of zeros for feet. 0'-6" = OFF   6" = ON
Inches: Controls the display of zeros for inches.   6'-0" = OFF   6" = ON

Angular Dimensions
Units Format: Sets the units format for Angular. Does not affect Linear.
Precision: Sets the number of decimal places past the whole degree.
Zero suppression: Same as Linear.

TRANS-SPATIAL DIMENSIONING

When using True Associative dimensioning, the dimension is actually attached to the object. If the object changes, the dimension changes. True Associative dimensioning is a very important and powerful tool within AutoCAD.

True Associative dimensioning gets even better because it can also be <u>trans-spatial</u>. Trans-spatial means that you have the ability to place a dimension in paper space while the object you are dimensioning is in model space. Even though the dimension is in paper space, it is actually attached to the object, in model space.

For example,
1. You draw a house in model space.
2. Now select the layout tab.
3. Cut a viewport so you can see the drawing of the house.
4. Go to model space, adjust the scale of the viewport (model space) and lock it.
5. Now go to paper space and dimension the house.

Why is Trans-spatial dimensioning so great?
If you dimension in Paper Space you do not have to be concerned with the Drawing Scale Factor.

SHOULD YOU DIMENSION IN PAPER SPACE OR MODEL SPACE?
It is possible to dimension in either Paper Space or Model Space. Personally, I dimension in either space depending on the situation. Here are some things to consider.

Paper Space dimensioning
Pro's
1. You never have to worry about the drawing scale factor. [Overall scale will remain set to 1.]
2. Dimensions do not change appearance if you adjust the scale of the viewport.
3. All dimensions will have the same appearance in all viewports.

Con's
1. Qdim is not trans-spatial. (You can only use it in model space.)
2. Dimensions sometimes temporarily float away from objects if you move the objects. (Use "Dimregen" to put them back in place.)

Model Space dimensioning
Pro's
1. Dimensions never float away from objects.
2. Qdim works in modelspace.

Con's
1. You must change the overall scale to match the drawing scale factor.
2. Dimensions will appear different in each viewport if the viewports have different adjusted scales.

Section 4
Drawing Entities

ARC

TYPING = A \<enter>
PULLDOWNS = DRAW / ARC
TOOLBARS =DRAW

There are 10 ways to draw an ARC in AutoCAD.   Not all of the ARCS options are easy to create so you may find it is often easier to **trim a Circle** or use the **Fillet** command.

On the job, you will probably only use 2 of these methods. Which 2 depends on the application.

An **ARC** is a segment of a circle and must be less than 360 degrees.

Most ARCS are drawn counter-clockwise but you will notice in the examples on the following pages that some may be drawn clockwise by entering a negative input.

Examples of each of the ARC options are shown on the following pages.

Also, refer to the "Help" menu for additional examples of the use of the Arc command.

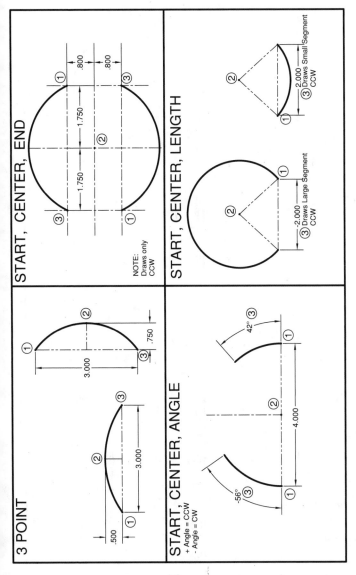

3 POINT

START, CENTER, END

START, CENTER, ANGLE
+ Angle = CCW
- Angle = CW

START, CENTER, LENGTH

NOTE:
Draws only
CCW

③ Draws Small Segment
CCW

③ Draws Large Segment
CCW

4-3

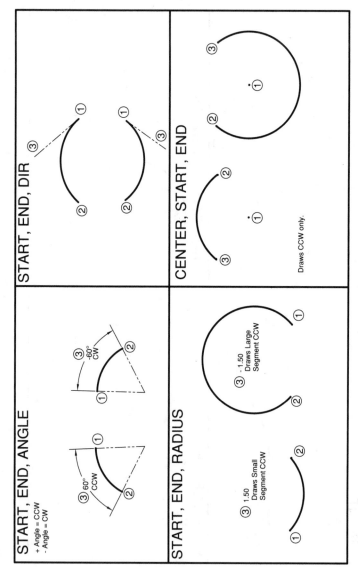

START, END, ANGLE
+ Angle = CCW
- Angle = CW

③ 60° CCW

③ -60° CW

START, END, DIR

START, END, RADIUS

③ 1.50
Draws Small
Segment CCW

③ -1.50
Draws Large
Segment CCW

CENTER, START, END

Draws CCW only.

4-4

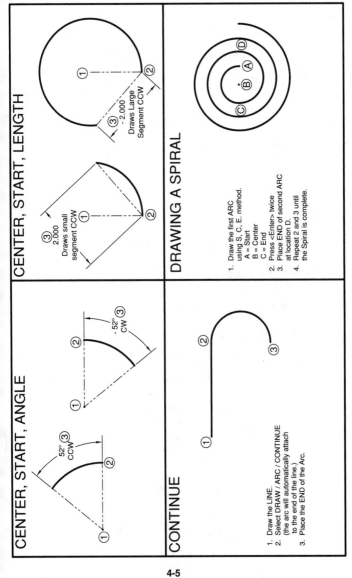

CENTER, START, LENGTH

Draws Large Segment CCW
-2.000

Draws small segment CCW
2.000

CENTER, START, ANGLE

52° CCW

-52° CW

DRAWING A SPIRAL

1. Draw the first ARC using S, C, E. method.
   A = Start
   B = Center
   C = End
2. Press <Enter> twice
3. Place END of second ARC at location D.
4. Repeat 2 and 3 until the Spiral is complete.

CONTINUE

1. Draw the LINE.
2. Select DRAW / ARC / CONTINUE (the arc will automatically attach to the end of the line.)
3. Place the END of the Arc.

BLOCKS

A **BLOCK** is a group of objects that have been converted into ONE object. A Symbol, such as a transistor, bathroom fixture, window, screw or tree, is a typical application for the block command. First a BLOCK must be created. Then it can be INSERTED into the drawing. An inserted Block uses less file space than a set of objects copied.

CREATING A BLOCK

1. First draw the objects that will be converted into a Block.

    For this example a circle and 2 lines are drawn.

2. Select the **BMAKE** command using one of the following:

    TYPE = B
    PULLDOWN = DRAW / BLOCK / MAKE
    TOOLBAR =DRAW

    (The Block Definition dialog box will appear.)

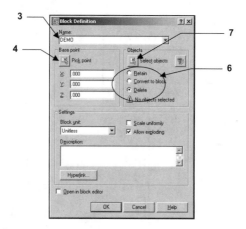

3. Enter the New Block name in the **Name** box.

4. Select the **Pick Point** button. (Or you may type the X, Y and Z coordinates.)
   The Block Definition box will disappear and you will return temporarily to the
   drawing.

5. Select the location where you would like the insertion point for the Block.
   Later when you insert this block, the block will appear on the screen attached to the
   cursor at this insertion point. Usually this point is the CENTER, MIDPOINT or
   ENDPOINT of an object.

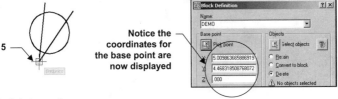

**Notice the
coordinates for
the base point are
now displayed**

6. Select an option.

   It is important that you select one and understand the options below.

   Retain
   If this option is selected, the original objects will stay visible on the screen after the
   block has been created.

   Convert to block
   If this option is selected, the original objects will disappear after the block has been
   created, but will immediately reappear as a block. It happens so fast you won't even
   notice the original objects disappeared.

   Delete
   If this option is selected, the original objects will disappear from the screen after the
   block has been created.

7. Select the **Select Objects** button.

   The Block Definition box will disappear and you will return temporarily to the
   drawing.

8. Select the objects you want in the block, then press <enter>.

The Block Definition box will reappear and the objects you selected should be illustrated in the Preview Icon area.

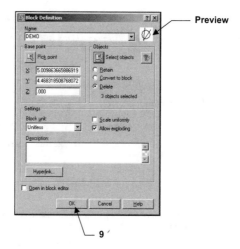

Preview

9

9. Select the **OK** button.
    The new block is now stored in the drawing's block definition table.

10. To verify the creation of this Block, select **Insert / Block**, select the Name (▼).
    A list of all the blocks, in this drawing, will appear.

ADDITIONAL DEFINITIONS OF OPTIONS

Block Units
You may define the units of measurement for the block. This option is used with the "Design Center" to drag and drop with Autoscaling. The Design Center is an advanced option and is not discussed in this book.

Scale Uniformly
Specifies whether or not the block is prevented from being scaled non-uniformly during insertion.

Allow Exploding
Specifies whether or not the block can be exploded after insertion.

Description
You may enter a text description of the block.

Hyperlink
Opens the **insert Hyperlink dialog box** which you can use to associate a hyperlink with the block.

HOW LAYERS EFFECT BLOCKS

If a block is created on Layer 0:

1. When the block is inserted, it will take on the properties of the current layer.
2. The inserted block will reside on the layer that was current at the time of insertion.
3. If you Freeze or turn Off the layer the block was inserted onto, the block will disappear.
4. If the Block is **Exploded**, the objects included in the block will revert to their original properties of layer 0.

If a block is created on Specific layers:

1. When the block is inserted, it will retain its own properties. It **will not** take on the properties of the current layer.
2. The inserted block **will reside** on the current layer at the time of insertion.
3. If you **freeze** the layer that was current at the time of insertion the block <u>will</u> disappear.
4. If you turn **off** the layer that was current at the time of insertion the block <u>will not</u> disappear.
5. If you **freeze** or turn **off** the blocks original layers the block <u>will</u> disappear.
6. If the Block is **Exploded**, the objects included in the block will go back to their original layer.

INSERTING BLOCKS

A **BLOCK** can be inserted at any location on the drawing. When inserting a Block you can **SCALE** or **ROTATE** it.

1. Select the INSERT command using one of the following:

    TYPE = DDINSERT
    PULLDOWN = INSERT / BLOCK
    TOOLBAR =DRAW

    The INSERT dialog box will appear.

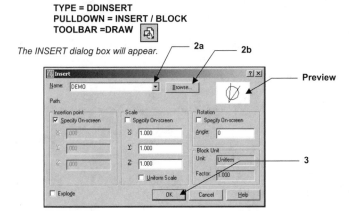

2. Select the **BLOCK** name.
    a. If the block is already in the drawing that is open on the screen, you may select the block from the drop down list shown above
    b. If you want to insert an entire drawing, select the Browse button to locate the drawing file.

3. Select the **OK** button.

    This returns you to the drawing and the selected block should be attached to the cursor.

4. Select the insertion location for the block by pressing the left mouse button or typing coordinates.

    Command: _insert
    Specify insertion point or **[Scale/X/Y/Z/Rotate/PScale/PX/PY/PZ/PRotate]:**

NOTE: If you want to scale or rotate the block before you actually place the block, press the right hand mouse button and you may select an option from the menu or select an option from the command line menu shown above.

    You may also "preset" the insertion point, scale or rotation.

PRESETTING THE <u>INSERTION POINT</u>, <u>SCALE</u> or <u>ROTATION</u>

You may preset the **Insertion point, Scale or Rotation** in the <u>INSERT</u> box instead of at the command line.

1. Remove the check mark from any of the **"Specify On-screen"** boxes.
2. Fill in the appropriate information describe below:

> **Insertion point**
> Type the X and Y coordinates <u>from the Origin</u>. The Z is for 3D only.
> *The example below indicates the block's insertion location will be 5 inches in the X direction and 3 inches in the Y direction, <u>from the Origin</u>.*
>
> **Scale**
> You may scale the block proportionately by typing the scale factor in the X box and then check the <u>Uniform Scale box</u>.
> If the block will be scaled non-proportionately, type the different scale factors in both X and Y boxes.
> *The example below indicates that the block will be scale proportionate at a factor of 2.*
>
> **Rotation**
> Type the desired rotation angle relative to its current rotation angle.
> *The example below indicates the block will be rotated 45 degrees from its originally created angle orientation.*

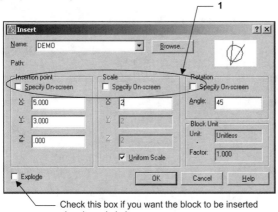

Check this box if you want the block to be inserted already exploded.

PURGE UNWANTED BLOCKS

How to delete unwanted blocks.

1. Select **File / Drawing Utilities / Purge**
2. Select the **+** sign beside **Blocks**
3. Select the block that you wish to delete.
4. Select **Purge** button.
5. Select **Close** button.

Note: You can't purge a block that is in use
within the drawing.

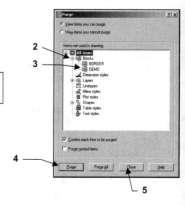

WHERE ARE BLOCKS SAVED?

When you create a Block it is saved <u>within the drawing you created it in</u>. (If you open another drawing you will not find that block.)

BLOCK EDITING – Versions 2004 & 2005

f you would like to add, delete or change objects within an existing block, you may do it easily with the command **Refedit**.

Refedit allows you to make changes to a block, saves the changes and updates all ▪ther previously inserted blocks within the drawing automatically.
The changes only affect the current drawing.

1. Select the Refedit command using one of the following:

> **TYPE = Refedit**
> **PULLDOWN = Tools / Xref & Block In-Place Editing / Edit Ref In-Place**
> **TOOLBAR = MODIFY II**

2. Select the Block you wish to edit. (Click on it, can't use a window)

The Reference Edit dialog box should appear.

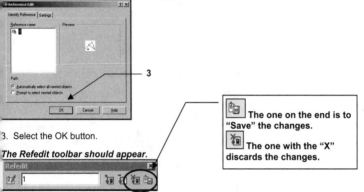

3. Select the OK button.

> The one on the end is to "Save" the changes.
>
> The one with the "X" discards the changes.

The Refedit toolbar should appear.

Note: The selected block will remain bold but all other blocks of the same name will fade to gray. This is to emphasis the selected block.

4. Make the changes to the selected block.

Note: As long as the "Refedit" toolbar is open, anything you add, erase or change will affect the selected block.

5. Select the "**Save**" icon on the far right end of the toolbar.

6. A warning will appear. Select the OK button.

The block has been redefined and all of the existing blocks have been updated to reflect the changes you made. (These changes affect the current drawing only)

BLOCK EDITING – Versions 2006 & 2007

How to change the design or insertion point of a block previously inserted.

1. Double click on the Block that you wish to change.

   *The "**Edit Block Definition**" dialog box will appear.*

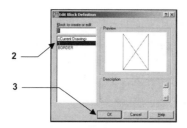

2. Select the name of the Block that you wish to change.

3. Select the **OK** button.

4. The "**Do you want to...**" dialog box appears. **Select <u>NO</u> for now**.

5. The Block that you selected should appear large on the screen. <u>Make the changes</u>.

6. Select the **Close Block Editor** button. (Located at the top of the screen on the tool bar.)

   *The **Save Changes** box will appear.*

7. Select **Yes**. You will be returned to the drawing and **ALL** of the blocks with the same name will be updated with the changes that you made.

DYNAMIC BLOCKS

What are Dynamic Blocks?

Dynamic Blocks are blocks that can be dynamically modified to represent different variations of the block.

An example could be a window or a screw. You will probably have many different sizes of windows in a floorplan or length of screws in an assembly. Typically you would create a block for every variation. This would increase the size of your symbols library.

Dynamic blocks allow you to create only one block but then define different lengths, size, rotation etc.

Although it is a little difficult to master and you may or may not find it necessary to use.

What makes a Block Dynamic?

Blocks are made dynamic by adding <u>Parameters</u> and <u>Actions</u>. The Block Editor is used to add these features. Once these features are added, the block can be changed easily after inserting. You will modify the blocks appearance using the custom grips or properties.

Previous to the Block Editor, blocks with many variations, built in, had to be created using AutoCAD's programming code AutoLisp. Those of us that are not programming types found this a bit difficult and would prefer to create all of the variations by drawing them and adding them to the symbol library. But now AutoCAD has made it possible for non-programming types to create Dynamic Blocks. Although I have to admit that it is not a piece of cake. It takes some studying to master this feature. It still has a programming feel.

The following is the Block Authoring Palettes, Parameters, Actions and Parameter sets.

DEFINING A DYNAMIC BLOCK – Version 2006 only

Let's try creating a very simple Dynamic Block. If you find this too difficult, Dynamic Blocks are not for you. If you find it fascinating and can visualize great advantages by using this new feature you will want to learn more.

1. Open **My Feet-inches setup**.

2. Select the **24 X 18 (Qtr-ft)** layout tab.

3. Draw the window shown below. Do not dimension.

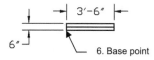

6. Base point

4. Select the **Make Block** command.

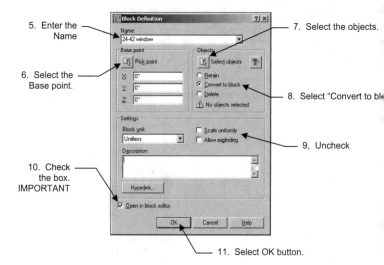

5. Enter the Name

6. Select the Base point.

7. Select the objects.

8. Select "Convert to bl‹

9. Uncheck

10. Check the box. IMPORTANT

11. Select OK button.

*Note: The screen has changed. You have now entered the **BLOCK EDITOR**. The Block Editor Toolbar, Block Authoring Palettes and the 24-42 Window appear.*

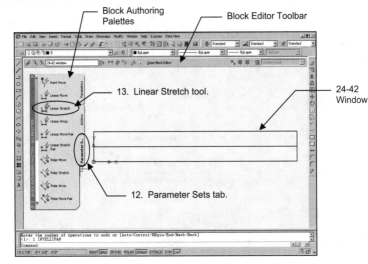

Block Authoring Palettes

Block Editor Toolbar

13. Linear Stretch tool.

24-42 Window

12. Parameter Sets tab.

12. Select the **Parameter Sets** tab.

13. Select the **Linear Stretch** tool.

14. Select "**Value Set**" option from the command line.

15. Select "**Increment**" option from the command line.

16. Enter the following values:
Increment distance: **3**
Minimum distance: **24**
Maximum distance: **42**

17. Place the parameter Start and End points:

Start ——

—— End

4-17

18. Place the **Distance parameter** label.

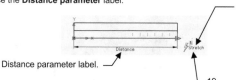

Notice the **Stretch** labe
The **exclamation** symb
indicates that the "actio
has not been defined ye

Distance parameter label. —

— 19

19. Double click on the "Stretch Action" label.

20. Define the end to stretch by drawing a **crossing window**.

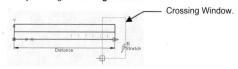

Crossing Window.

21. Draw the window again to select the objects.
   (This seems redundant but it is required.)

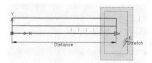

22. Press **<enter>** to stop selecting objects.
   Notice the exclamation point is gone and the increments are visible.

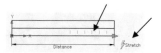

23. Select the "Save Block Definition" button.

24. Select the "Close Block Editor".

24-42 window   Close Block Editor

— 23   — 24

25. Click on the Window block.
    Notice the grips appear.

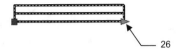

— 26

26. Click on the "**Block Stretch** " grip.

    You may slide the cursor to the left in 3" increments to resize the window from 42"
    to 24".

Dynamic Blocks take a lot of time to master.
The above example was merely to introduce you to Dynamic Blocks.
There are many more tools that you may find useful.
Experiment with some of the other tools. Refer to the AutoCAD Help menu for
more explanation.

DEFINING A DYNAMIC BLOCK- Version 2007 0nly

Let's try creating a very simple Dynamic Block. If you find this too difficult, Dynamic
Blocks are not for you. If you find it fascinating and can visualize great advantages by
using this new feature you will want to learn more.
The instructions below will show you how to add dynamic stretch to an existing block.

First you need to create a block or insert an existing block.

1. Open **My Feet-inches setup**.

2. Select the **24 X 18 (Qtr-ft)** layout tab.

3. Draw the window shown below. Do not dimension.

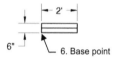

    6"   6. Base point

4. Select the **Make Block** command.

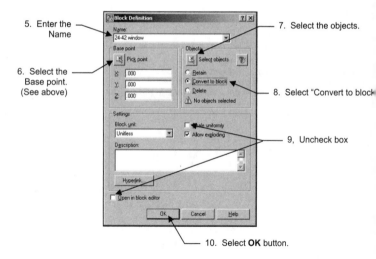

5. Enter the Name

6. Select the Base point.
(See above)

7. Select the objects.

8. Select "Convert to block

9. Uncheck box

10. Select **OK** button.

Next you have to define the Parameter.

11. Select **Tools / Block Editor** or type **BEDIT** on the command line.

The **Edit Block Definition** dialog box should appear.

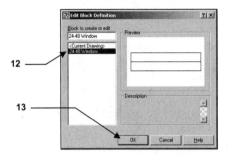

12. Select the block.

13. Select the **OK** button.

Note: Click on *NO* to view AutoCAD's demo on *"How to define a dynamic block"*
You can always see it later.

The **Block Authoring Palettes** and the **Block Editor** tool bar will appear.

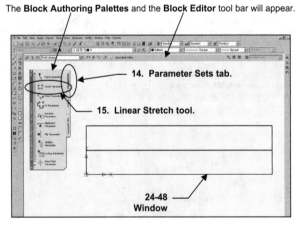

14. Select the **Parameters** tab.

15. Select the **Linear** parameter tool.

16. Select "**Value Set**" option from the command line by typing **V <enter>.**

17. Select "**Increment**" option from the command line by typing **I <enter>.**

18. Enter the following values:
    Distance Increment : **3**
    Minimum distance: **24**
    Maximum distance: **48**

19. Specify **Start** and **End** points by snapping to the corners shown below.

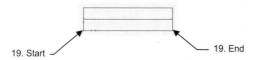

19. Start —   — 19. End

20. Specify Label location.

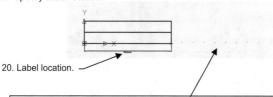

20. Label location. —

> *Notice the 8 little bars on the right hand end of the window. Those are the 3"
> incremental distances you defined in #18 above. When you have completed the
> following steps, you will be able to stretch the block from 24" to 48".*

21. Change the parameter to have only 1 grip.
    a. Click on the **Parameter**. (Distance)
    b. Right click and select **Grip Display** then **1**.

Now you need to assign the Action to the Parameter.

22. Select the **Action tab**.

23. Select the **Stretch Action** tool.

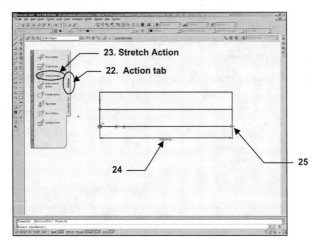

24. Select the **Parameter** to assign the Action. (Click on Distance)

25. Click on the **Grip**.

26. Define the **stretch frame** using a **crossing window**.

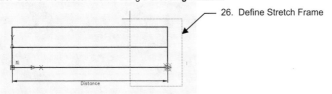

26. Define Stretch Frame

27. **Select the objects** to stretch using a **crossing window** again.
    Press <enter> to stop "selecting objects".

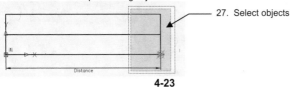

27. Select objects

4-23

28. Place the **Action label**.

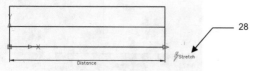

29. Select the "**Save Block Definition**" button.

30. Select the "**Close Block Editor**".

31. Click on the **24-48 Window** block.
    Notice the grips appear.

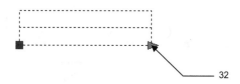

32. Click on the "**Block Stretch** " grip.

You may slide the cursor to the right in 3" increments to resize the window from 24"
to 48".

CENTERMARK

CENTERMARKS can ONLY be drawn with circular objects like Circles and Arcs.
You set the size and type.

The Center Mark has three types, **None, Mark** and **Line** as shown below.

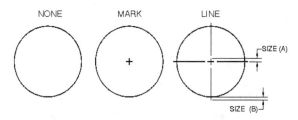

What does "SIZE" mean?

The **size** setting determines both, (A) the length of half of the intersection line and
(B) the length extending beyond the circle. (See above)

Where do you set the CENTERMARK "TYPE" and "SIZE"

1. Select the *Dimension Style* command.
2. Select: *New, Modify, or Override.*
3. Select: *(the tab shown)*
4. Select the *Center mark type*
5. Set the **Size**.

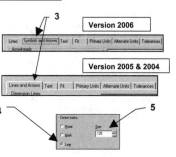

To draw a CENTER MARK

1. Select the **Center mark** command using one of the following:

> **TYPING = DIMCENTER or DCE**
> **PULLDOWN = DIMENSION / CENTERMARK**
> **TOOLBAR = DIMENSION**

2. Select arc or circle: *select the arc or circle with the cursor.*

CIRCLE

*There are 6 options to create a circle. The default option is "**Center, radius**".*
(Probably because that is the most common method of creating a circle.)
*We will try the "**Center, radius**" option first.*

1. Start the **Circle** command by using one of the following:
   TYPING = C <enter>
   PULLDOWN = DRAW / CIRCLE / Center, Radius
   TOOLBAR = DRAW

2. The following will appear on the command line:
   Command: _circle Specify center point for circle or [3P/2P/Ttr (tan tan radius)]:

3. Locate the center point for the circle by moving the cursor to the desired location in the drawing area and press the left mouse button.

4. Now move the cursor away from the center point and you should see a circle forming.

5. When it is approximately the size desired, press the left mouse button, or if you want the exact size, type the radius and then press <enter>.

Note: To use one of the other methods described below, first select the Circle command, then press the right mouse button. A "short cut" menu will appear. Select the method desired by placing the cursor on the option and pressing the left mouse button. Or you can type 3P or 2P or T, then press <enter>. (The short cut menu is simple and more efficient.)

Center, Radius: (Default option)
   1. Specify the center (P1) location.
   2. Specify the Radius (P2).

Center, Diameter:
   1. Specify the center (P1) location.
   2. Select the Diameter option using the shortcut menu or type "D" <enter>.
   3. Specify the Diameter (P2).

2 Points:

1. Select the 2 point option using the short cut menu or type 2P <enter>.
2. Specify the 2 points (P1 and P2) that will determine the Diameter .

3 Points:

1. Select the 3 Point option using the short cut menu or type 3P <enter>.
2. Specify the 3 points (P1, P2 and P3) on the circumference.
   The Circle will pass through all three points.

Tangent, Tangent, Radius:

1. Select the Tangent, Tangent, Radius option using the short cut menu or type T <enter>.

2. Select two objects (P1 and P2) for the Circle to be tangent to by placing the cursor on the object and pressing the left mouse button

3. Specify the radius.

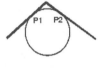

Tangent, Tangent, Tangent:

1. Select the Tangent, Tangent, Tangent option using the pull down menu. Note: May only be selected using the pull down menu. It is not available in the short cut menu or the command line.

2. Specify three objects (P1, P2 and P3) for the Circle to be tangent to by placing the cursor on the object and pressing the left mouse button. (The diameter will be calculated by the computer.)

DONUT

A Donut is a circle with *width*. You will define the **Inside** and **Outside** diameters.

1. Select the **DONUT** command using one of the following:

   TYPING =DO<enter>
   PULLDOWN = DRAW / DONUT
   TOOLBAR = DRAW

2. The following prompts will appear on the command line:

   Command: _donut
   Specify inside diameter of donut: *type the inside diameter*
   Specify outside diameter of donut: *type the outside diameter*
   Specify center of donut or <exit>: *place the center of the first donut*
   Specify center of donut or <exit>: *place the center of the second donut or*
                                       <enter> to stop

   Controlling the "FILL MODE"

   1. Command: *type FILL <enter>*

   2. Enter mode [ON / OFF] <OFF>: *type ON or OFF <enter>*

FILL = ON

FILL = OFF

3. Select **VIEW / REGEN** or type *REGEN <enter>* at the command line to update the drawing to the latest changes to the *FILL* mode.

ELLIPSE

An Ellipse may be drawn by specifying the 3 points of the axes or by defining the center point and major / minor axis points.

AXIS END METHOD
1. Select the **ELLIPSE** command using one of the following:

> **TYPING =EL <enter>**
> **PULLDOWN = DRAW / ELLIPSE**
> **TOOLBAR = DRAW**

2. The following prompts will appear on the command line:

Command: _ellipse
Specify axis endpoint of ellipse or [Arc/Center]: *place the first point of either the major or minor axis (P1).*
Specify other endpoint of axis: *place the other point of the first axis (P2)*
Specify distance to other axis or [Rotation]: *place the point perpendicular to the first axis (P3).*

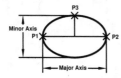

CENTER METHOD
1. Select the **ELLIPSE** command using one of the following:

> **TYPING =EL**
> **PULLDOWN = DRAW / ELLIPSE**
> **TOOLBAR = DRAW**

2. The following prompts will appear on the command line:

Command: _ellipse
Specify axis endpoint of ellipse or [Arc/Center]: *type C <enter>*
Specify center of ellipse: *place center of ellipse (P1)*
Specify endpoint of axis: *place first axis endpoint (either axis) (P2)*
Specify distance to other axis or [Rotation]: *place the endpoint perpendicular to the first axis (P3)*

GRADIENT FILLS

Gradients are fills that gradually change from dark to light or from one color to another. Gradient fills can be used to enhance presentation drawings, giving the appearance of light reflecting on an object, or creating interesting backgrounds for illustrations.

Gradients are definitely fun to experiment with but you will have to practice a lot to achieve complete control.

1. Select the BHATCH command using one of the following:

> **TYPE = BH**
> **PULLDOWN = DRAW / HATCH**
> **TOOLBAR = DRAW** 🔲
>
> The following dialog box appears:

2. Select the Gradient tab.
Note: Gradient tab not available in version "LT".

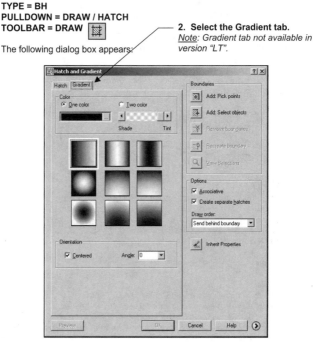

2. Select the Gradient tab.
3. Select the area for the gradient fill. (Pick Points or Select Objects)
4. Choose the Gradient settings. (Refer to the following page)
5. Preview
6. Accept or make changes.

GRADIENT FILLS continued...

ONE COLOR
Click the ... button to the right of the color swatch
to open the Select Color dialog box.
Choose the color you want.

Use the "**Shade and Tint**" slider to choose the
gradient range from lighter to darker.

TWO COLOR
Click each ... button to choose a color.
When you choose two colors the transition is both
from light to dark and from the first color to the
second.

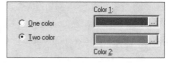

GRADIENT STYLES
Select one of the 9 gradient styles. (The selected
style will have a white box around it.)

CENTER
Select "**Centered**", with a check mark in the box,
to create a symmetrical fill.
Remove the check mark to move the "highlight"
up and to the left.

ANGLE
Specify an angle for the "highlighted" area from
the drop down list.
The angle rotates counter clockwise, 0 is upper
left corner.)

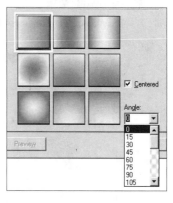

NOTES:
1. When you create a gradient, it appears in front of the
boundary and sometimes obscures the object's outline.
To send the gradient hatch behind the boundary, specify the
Draw Order.

2. It is good drawing management to always place gradient fill on it's own layer.

3. You may also make gradient fill appear or disappear with the **FILL** command.
Type "**FILL**" **<enter>** on the command line. Then type "**ON**" or "**OFF**".
Select **VIEW / REGEN** to see the effect.

HATCH – Version 2005 and 2004

The **BHATCH** command is used to create hatch lines for section views or filling areas with specific patterns.

To draw **hatch** you must start with a closed boundary. A closed boundary is an area completely enclosed by objects. A rectangle would be a closed boundary. You simply pick inside the closed boundary. BHATCH locates the area and automatically creates a temporary polyline around the outline of the hatch area. After the hatch lines are drawn in the area, the temporary polyline is automatically deleted.

Note: A Hatch set is one object. If you explode it, it will return to many objects.

1. Select the BHATCH command using one of the following:

TYPE = BH
PULLDOWN = DRAW / HATCH
TOOLBAR = DRAW

The following dialog box appears: — **Gradient tab** not
available in version "LT"

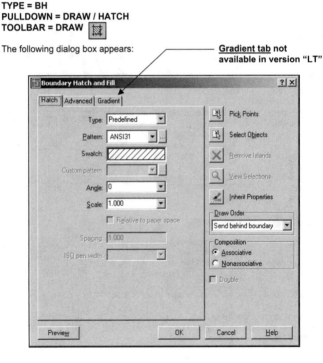

HATCH (continued)

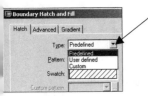

2. Select the hatch "TYPE"
Select one of the following:
PREDEFINED, USER DEFINED or CUSTOM
(Descriptions of each are listed below.)

a. *PREDEFINED*

AutoCAD has many predefined hatch patterns. These patterns are stored in the acad.pat and acadiso.pat files. (You may also purchase patterns from other software companies.)

> **Note:** Using Hatch patterns will greatly increase the size of the drawing file. So use them conservatively.

To select a pattern by name, click on the "Pattern" down arrow. A drop-down list of available patterns will appear.

To select a pattern by appearance, click on the (...) button. This will display the "Hatch Pattern Palette" dialog box.

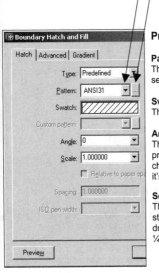

Predefined Pattern Properties

Pattern
This box displays the name of the Pattern you selected.

Swatch
The selected pattern is displayed here.

Angle
This determines the rotation angle of the pattern. A pre-designed pattern has a default angle of 0. If you change this Angle it will rotate the pattern relative to it's original design.

Scale
The value in this box is the scale factor. A good starting point would be to enter the scale factor of the drawing. Such as: If the drawing will be plotted at ¼" = 1, "4" is the scale factor.

HATCH (continued

b. *USER DEFINED*
This selection allows you to simply draw continuous lines. (No special pattern) You specify the Angle and the Spacing between the lines. (This selection does not increase the size of the drawing file like Predefined)

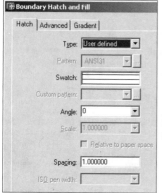

User-defined Pattern Properties

Swatch
A sample of the angle and spacing settings is displayed here.

Angle
Specify the actual angle of the hatch lines. (0 to 180)

Spacing
Specify the actual distance between each hatch line.

Double
If this box is checked, the hatch set will be drawn first at the angle specified, then a second hatch set will be drawn rotated 90 degrees relative to the first hatch set, creating a criss-cross affect.
(This option is only available when using "User Defined")

c. *CUSTOM*
See the AutoCAD Customization Guide for information on creating and saving custom hatch patterns.

3. Select the Composition Associative or Non Associative.

a. Associative: The hatch set is one entity and if the boundary size is changed the hatch will automatically change to the new boundary shape.

b. Non Associative: The hatch set is exploded into multiple entities and if the boundary shape is changed the hatch set will not change.

HATCH (continued)

4. Select the Area you want to Hatch using "Pick Points or Select Objects.

a. PICK POINTS (You will use this option primarily)
Select the **PICK POINTS** box then select a point inside the area you want to hatch. A boundary will automatically be determined.

b. SELECT OBJECTS
Select the boundary by selecting the object(s). The objects must form a closed shape with no gaps or overlaps.

Click inside the area

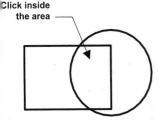

Gap Tolerance
If a warning appears stating "Valid hatch boundary not found" you probably have a gap in the boundary. If this occurs you can find the gap and fix it or you can set the Gap Tolerance to a value from 0 to 5000. (The default value is 0.) Any gaps equal to or smaller than the value you specify are ignored, and the boundary is treated as closed.
To set the gap tolerance select the "Advanced" tab. Gap tolerance box is located in the lower left corner.

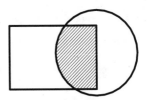

5. Preview the Hatch.
   a. After you have selected the boundary, press the right mouse button and select "**Preview**" from the short cut menu. This option allows you to preview the hatch set before it is actually applied to the drawing.
   b. If the preview is not what you expected, press the **ESC** key, make the changes and preview again.
   c. When you are satisfied, **right click** or press **<enter>** to accept.

Note: It is always a good idea to take the extra time to preview the hatch. It will actually save you time in the long run.

HATCH – Version 2007 & 2006

The **BHATCH** command is used to create hatch lines for section views or filling areas with specific patterns.

To draw **hatch** you must start with a closed boundary. A closed boundary is an area completely enclosed by objects. A rectangle would be a closed boundary. You simply pick inside the closed boundary. BHATCH locates the area and automatically creates a temporary polyline around the outline of the hatch area. After the hatch lines are drawn in the area, the temporary polyline is automatically deleted.

Note: A Hatch set is one object. If you explode it, it will return to many objects.

1. Select the BHATCH command using one of the following:

   TYPE = BH
   PULLDOWN = DRAW / HATCH
   TOOLBAR = DRAW

   The following dialog box appears: ———— **Gradient tab not available in version "LT"**

HATCH (continued)

. Select the hatch "TYPE"
Select one of the following:
PREDEFINED, USER DEFINED or CUSTOM
(Description of each is listed below.)

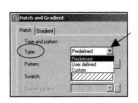

PREDEFINED
AutoCAD has many predefined hatch patterns. These patterns are stored in the acad.pat and acadiso.pat files. (You may also purchase patterns from other software companies.)

> **Note:** Using Hatch patterns will greatly increase the size of the drawing file.
> So use them conservatively.

To select a pattern by name, click on the "Pattern" down arrow. A drop-down list of available patterns will appear.

To select a pattern by appearance, click on the (…) button. This will display the "Hatch Pattern Palette" dialog box.

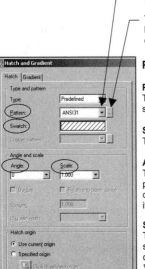

Predefined Pattern Properties

Pattern
This box displays the name of the Pattern you selected.

Swatch
The selected pattern is displayed here.

Angle
This determines the rotation angle of the pattern. A pre-designed pattern has a default angle of 0. If you change this Angle it will rotate the pattern relative to its original design.

Scale
The value in this box is the scale factor. A good starting point would be to enter the scale factor of the drawing. Such as: If the drawing will be plotted at ¼" = 1,   "4" is the scale factor.

4-37

HATCH (continued

b. _USER DEFINED_
This selection allows you to simply draw continuous lines. (No special pattern) You specify the Angle and the Spacing between the lines. (This selection does not increase the size of the drawing file like Predefined)

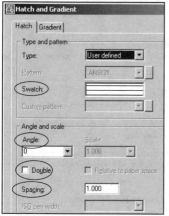

User defined Properties

Swatch
A sample of the angle and spacing settings is displayed here.

Angle
Specify the actual angle of the hatch lines. (0 to 180)

Spacing
Specify the actual distance between each hatch line.

Double
(This option is only available when using "User Defined")
If this box is checked, the hatch set will be drawn first at the angle specified, then a second hatch set will be drawn rotated 90 degrees relative to the first hatch set, creating a criss-cross affect.

3. Select the Options.

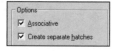

Associative: The hatch set is one entity and if the boundary size is changed the hatch will automatically change to the new boundary shape.

Create separate hatches: Controls whether HATCH creates a single hatch object or separate hatch objects when selecting several closed boundaries.

HATCH (continued)

4. Select the Area you want to Hatch using "Pick Points or Select Objects.

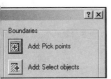

a. Add: Pick points (You will use this option primarily)
Select the **Pick points** box then select a point inside the area you want to hatch. A boundary will automatically be determined.

b. Add: Select objects
Select the boundary by selecting the object(s).

Click inside
the area

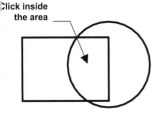

Gap Tolerance

If a warning appears stating "Valid hatch boundary not found" you probably have a gap in the boundary. If this occurs you can find the gap and fix it or you can set the Gap Tolerance to a value from 0 to 5000. (The default value is 0.) Any gaps equal to or smaller than the value you specify are ignored, and the boundary is treated as closed.

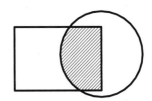

To set the gap tolerance: Select the "More Options" button.

 — More Options

5. Preview the Hatch.
a. After you have selected the boundary, press the right mouse button and select "**Preview**" from the short cut menu. This option allows you to preview the hatch set before it is actually applied to the drawing.
b. If the preview is not what you expected, press the **ESC** key, make the changes and preview again.
c. When you are satisfied, **right click** or press **<enter>** to accept.

Note: It is always a good idea to take the extra time to preview the hatch. It will actually save you time in the long run.

EDITING HATCH

HATCHEDIT allows you to edit an existing hatch pattern in the drawing.
You simply double click on the hatch pattern that you want to change and the hatch dialog box will appear. Make the changes, preview and accept.

1. Double click on the Hatch that you wish to edit, or select the Hatch Edit command using one of the following and then select the Hatch to edit:

TYPE = HE
PULLDOWN = MODIFY / OBJECT / HATCH
TOOLBAR = MODIFY II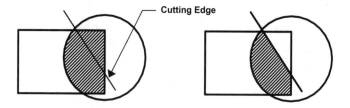

2. Make the changes to the settings and preview. Your changes should be displayed.

3. Right click to accept or ESC to make additional changes.

TRIMMING HATCH
You may trim a hatch set just like any other object. But... the hatch set will no longer be associative. Meaning, if you change the shape of the boundary the hatch set will not change.

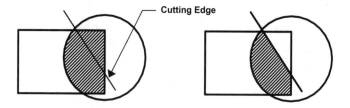

Cutting Edge

IGNORING HATCH OBJECTS

Occasionally, when you are dimensioning an object that has "Hatch Lines", your cursor will snap to the Hatch Line instead of the object that you want to dimension. To prevent this from occurring, select the option **"IGNORE HATCH OBJECTS"**.

EXAMPLE:

Ignore Hatch Objects
Not Selected

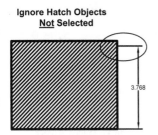

3.768

Ignore Hatch Objects
Selected

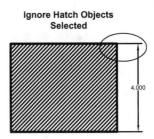

4.000

How to select the option "**IGNORE HATCH OBJECTS**".

1. Select **TOOLS / OPTIONS / DRAFTING**
2. Select "Ignore Hatch Objects" box.

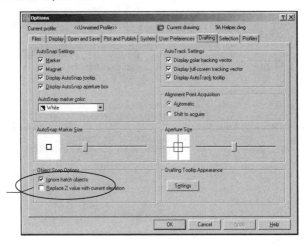

**Version LT
does not have
"Replace Z"
option**

DRAWING LINES

A **_LINE_** can be _one segment_ or a _series of connected segments_. Each segment is an individual object.

One segment   **Series of connected segments**

Start the Line command by using one of the following methods:

Type = L <enter>
PULLDOWN MENU = DRAW / LINE
TOOLBAR = DRAW

Lines are drawn by specifying the locations for each endpoint.
Move the cursor to the location of the **"first"** endpoint (1) then press the left mouse button. Move the cursor again to the **"next"** endpoint (2) and press the left mouse button. Continue locating **"next"** endpoints until you want to stop.

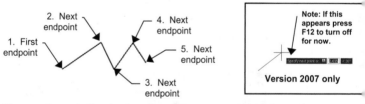

2. Next endpoint
4. Next endpoint
1. First endpoint
5. Next endpoint
3. Next endpoint

Note: If this appears press F12 to turn off for now.

Version 2007 only

There are 3 ways to **Stop** drawing a line: **1**. Press <enter> key. **2**. Press <Space Bar> or **3**. Press the right mouse button then select enter from the short cut menu.

To draw a perfectly **Horizontal** or **Vertical** line select the **ORTHO** mode by clicking on the **ORTHO button** on the **Status Bar** or pressing **F8**.
Note: Ortho can be temporarily turned off while drawing by holding down the "Shift key". Release the "Shift key to resume.

Ortho Mode
NAP GRID ORTHO POLAR OSN

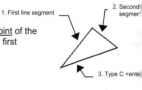

1. First line segment
2. Second segmer
3. Type C <ente

If you have drawn two or more line segments, the endpoint of the last line segment can be connected automatically to the first endpoint using the **CLOSE** option.

To use this option, draw two or more line segments, then type **C** <enter>.

POINT

A Point is an object that has no dimension and only has location. A Point may be represented by one of many Point styles shown below. Points are basically used to locate a point of reference. A location.

You may select the "Single" or "Multiple" point option. The Single option creates one point. The Multiple option will continue until you press the ESC key.

The only Object Snap mode that can be used with Point is "**Node**". (See Object Snap)

1. Select the **POINT** command using one of the following:

> **TYPING = PO <enter>**
> **PULLDOWN = DRAW / POINT (Single point or Multiple point)**
> **TOOLBAR = DRAW**

2. The following prompts will appear on the command line:

Command: _point
Current point modes: PDMODE=3 PDSIZE=0.000
Specify a point: *place the point location*
(If you selected "multiple point" you must press the "ESC" key to stop.)

TO SELECT A "POINT STYLE"

1. Select one of the following:
> **TYPE = DDPTYPE**
> **PULLDOWN = FORMAT / POINT STYLE**
> **TOOLBAR = NONE**

2. This dialog box will appear.

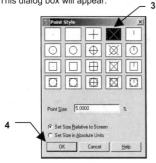

3. Select a tile.
4. Select the OK button.

POLYGON

A polygon is an object with multiple sides of equal length. You may specify from 3 to 1024 sides. A polygon appears to be multiple lines, but, in fact, it is one object. You can specify the edge length or the center and a radius. The radius can be drawn inside (Inscribed) an imaginary circle or outside (Circumscribed) an imaginary circle.

CENTER / RADIUS METHOD

1. Select the **Polygon** command using one of the following:

   TYPING = POL <enter>
   PULLDOWN = DRAW / POLYGON
   TOOLBAR = DRAW

2. The following prompts will appear on the command line:

   _polygon Enter number of sides <4>: *type number of sides <enter>*
   Specify center of polygon or [Edge]: *specify the center location*
   Enter an option [Inscribed in circle/Circumscribed about circle]<I>:*type I or C<enter>*
   Specify radius of circle: *type radius or locate with cursor.*

INSCRIBED   CIRCUMSCRIBED

EDGE METHOD

1. Select the **Polygon** command using one of the following:

   TYPING = POL
   PULLDOWN = DRAW / POLYGON
   TOOLBAR = DRAW

2. The following prompts will appear on the command line:

   _polygon Enter number of sides <4>: *type number of sides <enter>*
   Specify center of polygon or [Edge]: *type E <enter>*
   Specify first endpoint of edge: *place first endpoint of edge (P1)*
   Specify second endpoint of edge: *place second endpoint of edge (P2)*

POLYLINES

A **POLYLINE** is very similar to a LINE. It is created in the same way a line is drawn. It requires first and second endpoints. But a POLYLINE has additional features, as follows:

1. A **POLYLINE** is ONE object, even though it may have many segments.
2. You may specify a specific width to each segment.
3. You may specify a different width to the start and end of a polyline segment.

THE FOLLOWING ARE EXAMPLES OF POLYLINES WITH WIDTHS ASSIGNED.

Fill Mode = Off   Fill Mode = On

To turn FILL MODE on or off.

1. Command: *type FILL <enter>*
2. Enter mode [On / Off] <ON>: *type ON or Off <enter>*
3. Command: *type REGEN <enter> or select: View / Regen*

NOTE: *If you explode a POLYLINE it loses its width and turns into a regular line.*

THE FOLLOWING IS AN EXAMPLE OF DRAWING A POLYLINE WITH "WIDTH"

1. Select the POLYLINE command using one of the following:

    TYPE = PL
    PULLDOWN = DRAW / POLYLINE
    TOOLBAR = DRAW

    Command: _pline
2. Specify start point: *place the first endpoint of the line*
    Current line-width is 0.000
3. Specify next point or [Arc/Halfwidth/Length/Undo/Width]: *type w <enter>*
4. Specify starting width <0.000>: *type the desired width <enter>*
5. Specify ending width <0.000>: *type the desired width <enter>*
6. Specify next point or [Arc/Close/Halfwidth/Length/Undo/Width]: *place the next endpoint*
7. Specify next point or [Arc/Close/Halfwidth/Length/Undo/Width]: *place the next endpoint*
8. Specify next point or [Arc/Close/Halfwidth/Length/Undo/Width]: *place the next endpoint*
9. Specify next point or [Arc/Close/Halfwidth/Length/Undo/Width]: *type C <enter>*

OPTIONS:

WIDTH
Specify the starting and ending width.

You can create a tapered polyline by specifying different starting and ending widths.

HALFWIDTH
The same as Width except the starting and ending halfwidth specifies half the width rather than the entire width.

ARC
This option allows you to create a circular polyline less than 360 degrees.

CLOSE
The close option is the same as in the Line command. Close attaches the last segment to the first segment.

LENGTH
This option allows you to draw a polyline at the same angle as the last polyline drawn. This option is very similar to the OFFSET command. You specify the first endpoint and the length. The new polyline will automatically be drawn at the same angle as the previous polyline.

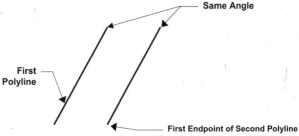

EDITING POLYLINES

The **POLYEDIT** command allows you to make changes to a polyline's option, such as the width. You can also change a regular line into a polyline and JOIN the segments.

1. Select the **POLYEDIT** command using one of the following:

> **TYPE = PE**
> **PULLDOWN = MODIFY / OBJECT/ POLYLINE**
> **TOOLBAR =MODIFY II**

Note: You may modify "Multiple" polylines simultaneously.

2. PEDIT Select polyline or [Multiple]: *select the polyline to be edited or "M"*
3. Enter an option [Close/Join/Width/Edit vertex/Fit/Spline/Decurve/Ltypegen/Undo]:
select an Option (descriptions of each are listed below.)

Note: If you select a line that is **NOT A POLYLINE**, the prompt will ask if you would like to turn it into a POLYLINE.

OPTIONS:

CLOSE
CLOSE connects the last segment with the first segment of an Open polyline. AutoCAD considers a polyline open unless you use the "Close" option to connect the segments originally.

OPEN
OPEN removes the closing segment, but only if the CLOSE option was used to close the polyline originally.

JOIN
The JOIN option allows you to join individual polyline segments into one polyline. The segments must have matching endpoints.

WIDTH
The WIDTH option allows you to change the width of the polyline. But the entire polyline will have the same width.

EDIT VERTEX
This option allows you to change the starting and ending width of each segment individually.

SPLINE
This option allows you to change straight polylines to curves.

DECURVE
This option removes the SPLINE curves and returns the polyline to its original straight line segments.

RECTANGLE

A Rectangle is a closed rectangular shape. It is one object not 4 lines.
You can specify the length, width, area, and rotation parameters.
You can also control the type of corners on the rectangle—fillet, chamfer, or square and the width of the Line.

First, let's start with a simple Rectangle using the mouse to select the corners.

1. Start the **RECTANGLE** command by using one of the following:

   TYPING = REC <enter>
   PULLDOWN = DRAW / RECTANGLE
   TOOLBAR = DRAW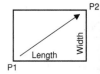

2. The following will appear on the command line:

Command: _rectang
Specify first corner point or [Chamfer/Elevation/Fillet/Thickness/Width]:

3. Specify the location of the first corner by moving the cursor to a location (**P1**) and then press the left mouse button.

   The following will appear on the command line:

Specify other corner point or [Area / Dimensions / Rotation]:

4. Specify the location of the **diagonal** corner (**P2**) by moving the cursor diagonally from the first corner (**P1**) and pressing the left mouse button.

   OR

   Type **D** <enter> (or press right mouse button and select "Dimensions" from the short cut menu)
   Specify length for rectangles <0.000>: *Type length <enter>.*
   Specify width for rectangles <0.000>: *Type width <enter>.*
   Specify other corner point or [Dimension]: *move the cursor up, down, right or left to specify where you want the second corner relative to the first corner and then press <enter> or press left mouse button.*

PARAMETERS – Rotation and Area
ROTATION - You may select the desired rotation angle <u>after</u> you place the first corner and <u>before</u> you place the second corner. The base point is the first corner. <u>Note: All new rectangles within the drawing will also be rotated unless you reset the rotation.</u>

AREA – You may define the size of the rectangle by inputting the Area and the length or width. Example: If you select Area and length, AutoCAD calculates the width

OPTIONS:
You may also preset the rectangle corners to angled or rounded and adjust the line
width using the Chamfer, Fillet and Width options.

CHAMFER
A chamfer is an angled corner. The Chamfer option automatically draws all 4 corners
with chamfers (all the same size). You must specify the distance for each side of the
corner as distance 1 and distance 2.

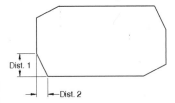

FILLET
A fillet is a rounded corner. The fillet option automatically draws all 4 corners with fillets
(all the same size). You must specify the radius for the rounded corners.

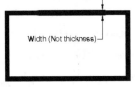

WIDTH
Sets the width of the rectangle lines. (Note: Do not confuse this with the Length and
Width. This makes the lines appear to have width.)

**Note: If you set the Chamfer, Fillet or Width to a value greater than "0", any new
rectangles will be affected until you reset the option to "0".**

ELEVATION and THICKNESS: Used in 3D only.

REVISION CLOUD STYLE

You may select one of 2 styles for the Revision Cloud; **Normal** or **Calligraphy**.
Normal will draw the cloud with one line width.
Calligraphy will draw the cloud with variable line widths to appear as though you used a chiseled calligraphy pen.

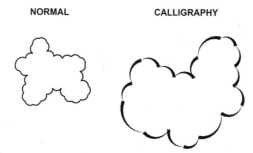

| NORMAL | CALLIGRAPHY |

1. Select the Revision Cloud command using one of the following:

   **TYPE = REVCLOUD
   PULLDOWN = DRAW / REVISION CLOUD
   PULLDOWN (LT) = TOOLS / REVISION CLOUD
   TOOLBAR = DRAW**

   Command: _revcloud
   Minimum arc length: .50   Maximum arc length: 1.00   Style: Normal

2. Specify start point or [Arc length/Object/Style] <Object>: *Select "Style"<enter>*

3. Select arc style [Normal/Calligraphy] <Calligraphy>:*Select "N or C"<enter>*

4. Specify start point or [Arc length/Object/Style] <Object>: *Select "Arc length"*

5. Specify minimum length of arc <.50>: *Specify the minimum arc length*

6. Specify maximum length of arc <1.00>: *Specify the maximum arc length*

7. Specify start point or [Object] <Object>: *Place cursor at start location & left click.*

8. Guide crosshairs along cloud path...*Move the cursor to create the cloud outline.*

9. Revision cloud finished. *When the cursor approaches the start point, the cloud closes automatically.*

CREATING A REVISION CLOUD

When you make a revision to a drawing it is sometimes helpful to highlight the revision for someone viewing the drawing. A common method to highlight the area is to draw a "Revision Cloud" around the revised area. This can be accomplished easily with the "Revision Cloud" command.

The Revision Cloud command creates a series of sequential arcs to form a cloud-shaped object. You set the minimum and maximum arc lengths. (Maximum arc length cannot exceed three times the minimum arc length. Example: Min = 1, Max can be 3 or less) If you set the minimum and maximum different lengths the arcs will vary in size and will display an irregular appearance.

Min & Max same length   **Min & Max different length**

To draw a Revision Cloud you specify the start point with a left click then drag the cursor to form the outline. AutoCAD automatically draws the arcs. When the cursor gets very close to the start point, AutoCAD snaps the last arc to the first arc and closes the shape.

1. Select the Revision Cloud command using one of the following:

   TYPE = REVCLOUD
   PULLDOWN = DRAW / REVISION CLOUD
   PULLDOWN (LT) = TOOLS / REVISION CLOUD
   TOOLBAR = DRAW

   Command: _revcloud
   Minimum arc length: .50   Maximum arc length: .50   Style: Normal

2. Specify start point or [Arc length/Object/Style] <Object>: *Select "Arc length"*

3. Specify minimum length of arc <.50>: *Specify the minimum arc length*

4. Specify maximum length of arc <.50>: *Specify the maximum arc length*

5. Specify start point or [Arc length/Object/Style] <Object>: *Place cursor at start location & left click.*

6. Guide crosshairs along cloud path...*Move the cursor to create the cloud outline.*

7. Revision cloud finished. *When the cursor approaches the start point, the cloud closes automatically.*

CONVERT A CLOSED OBJECT INTO A REV CLOUD

You can convert a closed object, such as a circle, ellipse, rectangle or closed polyline to a revision cloud. The original object is deleted when it is converted.
(If you want the original object to remain, in addition to the new rev cloud, set the variable "delobj" to "0". The default setting is "1".)

1. Draw a closed object such as a circle.

2. Select the Revision Cloud command using one of the following:

   TYPE = REVCLOUD
   PULLDOWN = DRAW / REVISION CLOUD
   PULLDOWN (LT) = TOOLS / REVISION CLOUD
   TOOLBAR = DRAW

   Command: _revcloud
   Minimum arc length: .50   Maximum arc length: .50   Style: Normal

3. Specify start point or [Arc length/Object/Style] <Object>: ***Select "Arc length"***

4. Specify minimum length of arc <.50>: ***Specify the minimum arc length***

5. Specify maximum length of arc <.50>: ***Specify the maximum arc length***

6. Specify start point or [Arc length/Object/Style] <Object>: ***Select "Object".***

7. Select object: ***Select the object to convert***

8. Select object: Reverse direction [Yes/No] <No>: ***Select Yes or No***
   Revision cloud finished.

REVERSE DIRECTION

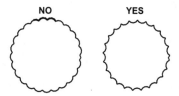

NO   YES

NOTE:
The Match Properties command will not match the arc length from the source cloud to the destination cloud.

Section 5
How to…

Add a Printer / Plotter

The following are step-by-step instructions on how to configure AutoCAD for your printer or plotter. These instructions assume you are a single system user. If you are networked or need more detailed information, please refer to your AutoCAD users guide.

Note: You can configure AutoCAD for multiple printers. Try adding the printer shown in this example and then add your own.

A. Select **File / Plotter Manager**
B. Select **"Add-a-Plotter"** Wizard

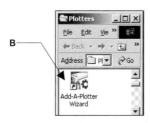

C. Select the **"Next"** button.

D. Select "**My Computer**" then **Next**.

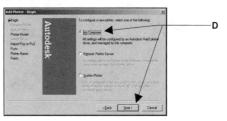

E. Select the **Manufacturer** and the specific **Model** desired then N**ext**.

(If you have a disk with the specific driver information, put the disk in the disk drive and select "Have disk" button then follow instructions.)

NOTE: Please
configure this
printer on
your system.
This **will not**
effect your
computer in
any negative
way.

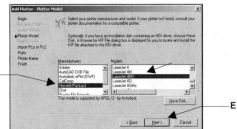

E

F. Select the **"Next"** box.

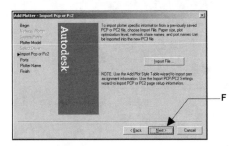

F

G. Select **"Plot to a port"**.
Then select **"Next"**.

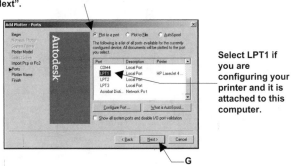

Select LPT1 if
you are
configuring your
printer and it is
attached to this
computer.

G

H. The Printer name, that you previously selected, should appear. Then select **"Next"**

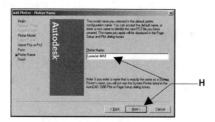

I. Select the **"Edit Plotter Configuration..."** box.

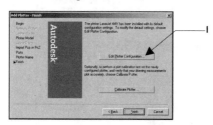

J. Select:
1. Device and Document Settings tab.
2. Media: Source and Size
3. Size: Ansi B (11 X 17 inches)
4. OK box.

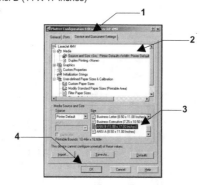

K. Select **"Finish"**.

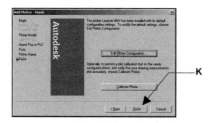

  K

L. Now check the **File / Plotter Manager**.

Is the printer / plotter there?

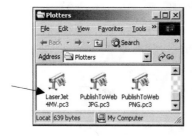

HOW TO CREATE A PAGE SET UP for Paperspace
Version 2004 only

*To setup a sheet in Paper Space (Layout tab), you must specify what <u>printing device</u> to use and what <u>paper size</u> to print on. These specifications are called the "**Page Setup**".*

SET UP THE PAPERSPACE ENVIRONMENT.
Setting up the Paperspace environment means name the layout tab, select the printer and specify paper size.

A. Select a **LAYOUT** tab.

Note: If the Page Setup dialog box shown below does not appear automatically, right click on the Layout tab, then select Page Setup.

    1. Type the new name for the Layout tab.

    2. Select the "Plot Device" tab.

    3. Select the Plotter / Printer.

    4. Select the Plot Style table (.ctb) from the list.

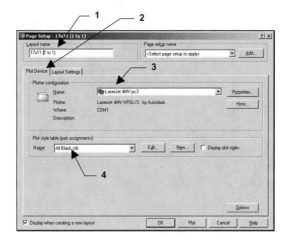

5. Select the "Layout Settings" tab.

6. Select the "Paper Size" from the list.

7. Select the "Scale".

8. Select "Layout". Note: You will be instructed later to select "Extents".

9. Select the "OK" button.

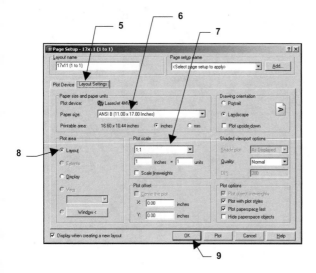

You should now have a sheet of paper displayed on the screen and the Layout tab should now be displayed with it's new name.

This sheet is in front of "Modelspace".

You may want to create a border, title and cut a viewport.

Draw the border, title block and notes in paperspace.
Cut a viewport to see through to modelspace.

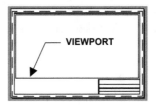

HOW TO CREATE A PAGE SET UP for PAPER SPACE
Version 2007, 2006 and 2005

To setup a sheet in Paper Space (Layout tab), you must specify what <u>printing device</u> to use and what <u>paper size</u> to print on. These specifications are called the "**Page Setup**".

SET UP THE PAPER SPACE ENVIRONMENT.
Setting up the Paper Space environment means selecting the printer and specifying paper size.

1. Select a **LAYOUT** tab.

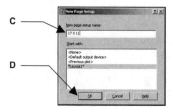

Note: If the "Page Setup Manager" dialog box shown below does not appear automatically, right click on the Layout tab, then select Page Setup Manager.

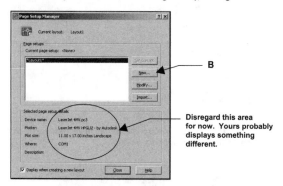

Disregard this area for now. Yours probably displays something different.

B. Select the **New** button. (*The "New Page Setup" dialog should appear*)

C. Type the new page setup name.

D. Select the **OK** button.

5-9

The "Page Setup" dialog box should appear.

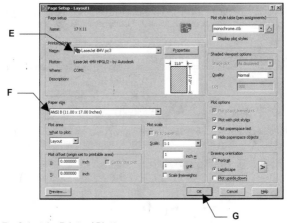

E. Select the **Printer / Plotter**.
F. Select the **Paper Size.**
G. Select the **OK** button.

The "Page Setup Manager" should have returned. Notice the new page setup name now appears in the list.

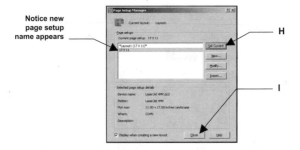

Notice new page setup name appears

H. Select the **Set Current** button.
I. Select the **Close** button.

You should now have a sheet of paper displayed on the screen.
This sheet is the size you specified in the "Page Setup".
This sheet is in front of the drawing that is in Model Space.
The dashed line represents the printing limits for the device that you selected.

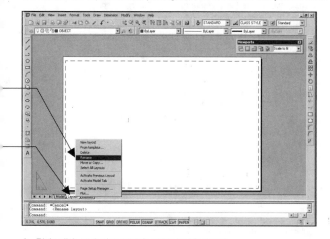

J. Right click on the Layout tab and select **Rename**.
K. Type the new name then select **OK**.

*Note: Renaming is optional but helpful. The name is usually a description of the
layout you just created. Example: 11 x 17 or Assembly or Floor Plan.*

ou may want to create a border, title and cut a viewport.
A. Draw the border, title block and notes in Paper Space.
B. Cut a viewport to see through to Model Space.

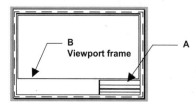

If you cut a viewport you will need to "adjust the scale" within the viewport.

Adjust Model Space Scale within the viewport.

A. Double click inside the viewport to change to model space.
   Note: The viewport <u>must</u> be unlocked to adjust the scale.

B. Select **View / Zoom / All** to make sure all of the drawing appears within the
   Viewport.

C. Open the Viewports toolbar.

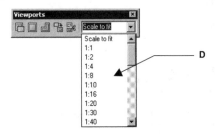

D. Select a scale from the drop down list.

Note: If you need to reposition your drawing within the viewport, use the **PAN** command
Do not use the Zoom commands. The zoom commands will change the scale that you
just adjusted.

E. Lock the Viewport

   Lock the Viewport.
   Now you may zoom as much as you desire and it will not affect the adjusted
   scale.

CREATE A TEMPLATE

1. Start AutoCAD, if you haven't already.

2. Select **File / Open**

3. Select the **Desktop** directory and locate the downloaded the files.

4. Select the file that you want to change to a template and then "**Open**" button.

Notice the 3 letter extension for a "drawing" file is ".dwg".

5. Select **"File / Save As..."**

6. Select the "**Files of type:**" down arrow ▾ to display different saving formats.
   Select "**AutoCAD Drawing Template (*.dwt)".**

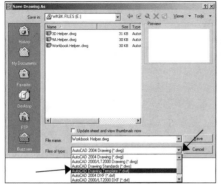

Notice the 3 letter extension for Template is ".dwt".

A list of all the AutoCAD templates will appear. (Note: Your list may be different)

7. Type the new name in the "**File name**:" box and then select the "**Save**" button.

Note: AutoCAD displays numerical first and then alphabetical.

Notice it was not necessary to type the extension .dwt because "Files of type" was previously selected.

8. Type a description and the select the "**OK**" button.

Now you have a template to use.

Using a template as a master setup drawing is good CAD management.

OPENING A TEMPLATE

To Open a Template follow the steps below.

1. Select **FILE / NEW**.

2. Select the **Use a Template** box (third from the left).

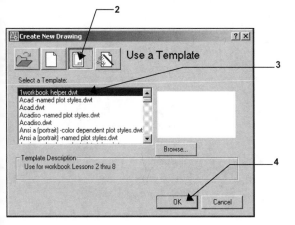

3. Select the template from the list of templates.

4. Select the **OK** button.

OPENING AN EXISTING DRAWING FILE

1. Start the command by using one of the following methods.

TYPING:	**OPEN <enter>** or press **CTRL + O**
PULLDOWN:	**FILE / OPEN**
TOOLBAR:	**STANDARD**

The Dialog box shown below should appear.

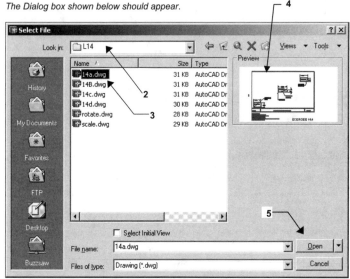

2. Select the **Drive and Directory** from the **"LOOK IN"** Box.

3. Select the drawing file from the list. (You may double click on the file name to automatically open the drawing)

4. The Preview window displays a "Thumbnail Preview Image".

5. Select the **OPEN** button.

EXITING AUTOCAD

1. Start the command by using one of the following methods:

TYPING: EXIT <enter> or   QUIT<enter>
PULLDOWN: FILE / EXIT (Safest method)
TOOLBAR: NONE

If any changes have been made to the drawing since the last save, the warning box below will appear asking if you want to **SAVE THE CHANGES**?

Select **YES, NO** or **CANCEL**.

VIEWPORTS

Viewports are only used in Paper Space (Layout tab).
Viewports are holes cut into the sheet of paper displayed on the screen.
Viewports are objects. They can be moved, stretched, scaled, copied and erased.
You can have multiple Viewports, and their size and shape can vary.

Note: It is considered good drawing management to create a layer for the Viewport "frames" to reside on. This will allow you to control them separately; such as setting the viewport layer to "No plot" so it will not be visible when plotting.

HOW TO CREATE A VIEWPORT

1. First, create your drawing in Model Space (Model tab) and save it.

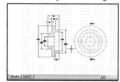

2. Select the "Layout1" tab. (If the "Page Setup Manager" dialog box appears, select the **CLOSE** button for now or refer to "How to Create a Page setup".)
3. You are now in Paper Space. Model Space appears to have disappeared, but a blank paper is actually in front of Model Space, preventing you from seeing your drawing. (Your Border, title block and notes will be drawn on this paper in Paper Space.)

4. Select layer "Viewport"
5. Open the "Viewports" toolbar

LT toolbar

6. Select the "Single Viewport" icon.

7. Draw the Viewport "frame" by specifying the location for the "first corner" and then the "opposite corner". (Similar to drawing a Rectangle, but **do not** use the Rectangle command.)

You should now be able to look through the Paper Space sheet to Model Space and see your drawing.

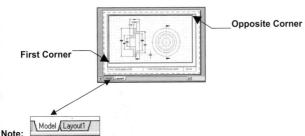

Opposite Corner

First Corner

Note:
You may go back to Model Space simply by clicking on the Model tab.
To return to Paper Space, click on the Layout tab.
(Make sure your grids are ON in Model Space and OFF in Paper Space. Otherwise you will have double grids)

EXAMPLE OF WHY WE WANT TO USE PAPER SPACE
I know you are probably wondering why you should bother with Paper Space. Paper Space is a great tool to manipulate your drawing for plotting.
Notice the drawing below. Viewport 1 displays the entire drawing. In Viewport 2 the scale has been adjusted to get a closer look at that section arrow. Viewport 3 not only has it's scale adjusted but the dimension layer is frozen (invisible) in the "<u>current viewport</u>" only. Notice the dimension layer is still thawed (visible) in the upper viewport. Can you guess which is the **Active** viewport at the moment? (That's right; Viewport 3)

Experiment and get familiar with Paper Space. It is a <u>very useful</u> tool.

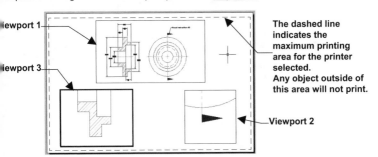

Viewport 1

Viewport 3

The dashed line indicates the maximum printing area for the printer selected. Any object outside of this area will not print.

Viewport 2

TO REACH THROUGH TO MODEL SPACE WHILE IN PAPER SPACE

Note: You may draw in model space (inside the viewport) while in Paper Space (Layout tab). You may switch to Model or Paper Space using one of the following methods.

Method 1.
Select the **Model / Paper** button on the Status Line.

Method 2.
If you double click <u>inside</u> the viewport frame, it will activate the Model Space.
The Model space UCS icon will appear within the viewport and the frame will appear heavy.

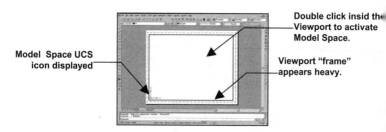

Model Space UCS icon displayed

Double click insid the Viewport to activate Model Space.

Viewport "frame" appears heavy.

Model Space Active

To return to Paper Space, double click in the gray area and the Paper Space icon will appear.

Double click in the gray area to activate Paper Space.

Paper Space Icon

Viewport "frame" appears thin.

Paper Space Active

VIEWPORTS TOOLBAR TOOLS

TOOLBAR SHOWN BELOW:

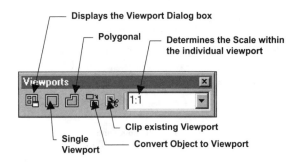

- Displays the Viewport Dialog box
- Polygonal
- Determines the Scale within the individual viewport
- Clip existing Viewport
- Convert Object to Viewport
- Single Viewport

TOOLBAR FOR <u>LT</u> SHOWN BELOW:

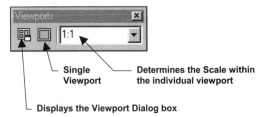

- Single Viewport
- Determines the Scale within the individual viewport
- Displays the Viewport Dialog box

HOW TO LOCK A VIEWPORT

After you have adjusted the scale within each viewport, you will want to "LOCK" the viewport so it will not change. <u>Here are 3 methods</u>.

Method 1. a. Make sure you are in Paper Space.
  b. Left click <u>once</u> on the Viewport frame
  c. Right Click (the short cut menu should appear)
  d. Select "**Display Locked**" - Select Yes or No.

Method 2. a. Left click <u>once</u> on the Viewport frame
  b. Right Click and select "Properties".
  c. Select **Misc / Displayed Locked**. (Select Yes or No)

Method 3. a. Type **MV** at the command line and press <enter>
  b. Select "Lock" <enter>
  b. Type On or Off <enter>
  c. Click on the Viewport frame.

Customizing your Wheel Mouse

A Wheel mouse has two or more buttons and a small wheel between
the two topside buttons. The default functions for the two top buttons
are as follows:
Left Hand button is for **input**
Right Hand button is for **Enter** or the **shortcut menu**.

You will learn more about this later. But for now follow the instructions below.

Using a Wheel Mouse with AutoCAD®

To get the most out of your Wheel Mouse set the **MBUTTONPAN** setting to **"1"** as
follows:

1. **At the command line,
type MBUTTONPAN <enter>**

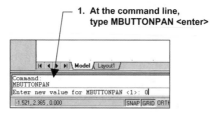

*After you understand the function of the "Mbuttonpan" variable, you can decide whether
you prefer the setting "0" or "1" as described below.*

MBUTTONPAN setting 0:

2. **Type: 0 or 1 <enter>**

ZOOM Rotate the wheel forward to zoom in
Rotate the wheel backward to zoom out

**OBJECT
SNAP** Object Snap menu will appear when you press the wheel

MBUTTONPAN setting 1: (Factory setting)

ZOOM Rotate the wheel forward to zoom in
Rotate the wheel backward to zoom out

**ZOOM
EXTENTS** Double click the wheel

PAN Press the wheel and drag

CALCULATING THE DRAWING SCALE FACTOR

Think about this....
When you adjust the scale within a Viewport all objects within the Viewport appear either larger or smaller. ALL OBJECTS, including <u>Text</u>, <u>Hatch</u> and <u>Dimension features</u> such as arrowheads. Even the spacing for non-continuous linetypes such as "dashed" appear larger or smaller. As a result, these objects may be too small or too large. So you must adjust the size of these objects.

If the objects appear smaller, you must increase the size of the Text height, Hatch scale or spacing and Dimension entities. If the objects appear larger, you must decrease the size of the Text height, Hatch scale or spacing and Dimension entities. I guess you figured that much out on your own.

But how do you know what size to make them?

Determining the size
To determine how much to increase or decrease you need to calculate the **drawing scale factor**. The drawing scale factor (DSF) means "<u>How many times smaller or larger did the drawing get when you adjusted the scale</u>".

Here are a few of the most commonly used scales and their <u>drawing scale factors</u>:

Scale	DSF	Scale	DSF
⅛" = 1'	12	1 : 2	2
¼" = 1'	48	2 : 1	1/2
⅛" = 1'	96	4 : 1	1/4

HOW TO CALCULATE THE DRAWING SCALE FACTOR (DSF):

The drawing scale factor is the reciprocal of the adjusted scale.

For example, if the scale has been adjusted to: 1/4" = 1'
Calculate the scale factor as follows:
    Adjusted scale: ¼" = 1'
1. Convert to decimals: .25 = 12
2. Divide 12 by .25 (12 ÷ .25 = 48)   (The Drawing scale factor is 48)

This means that the text height, hatch scale/spacing and dimension scale must be adjusted by 48.

Another example: if the scale has been adjusted to 4 : 1
    Adjusted scale: 4 : 1
1. Convert to decimals: 4 = 1
    (if necessary)
2. Divide 1 by 4   (1 ÷ 4 = 1/4)   (The Drawing scale factor is 1/4 or .25)

This means that text height, hatch scale/spacing and dimension scale must be adjusted by 1/4.

HOW THE DSF EFFECTS DIMENSIONING

Note: the following only affects dimensions in model space only.

When you adjust the scale within a Viewport, dimension features, such as arrowheads, will increase or decrease in appearance also.

Do not misunderstand. It will not affect the dimension "value", only the size of the dimension features, such as arrowheads.

<u>All you need to do is set the Dimension "overall scale" box to the Drawing Scale Factor.</u> Dimension feature sizes should remain as you normally have them set. Then AutoCAD will automatically adjust the size of all the dimension features as you draw the dimensions in the viewport.

HOW TO SET THE "OVERALL SCALE".

1. Select "**Format / Dimension Style**"
2. Select the dimension style that you want to change.
3. Select the "**Modify**" button.
4. Select the "**Fit**" tab.
5. Select "**Use overall scale of:**"
6. Enter the **DSF** in the box.
7. Select **OK** and close the Dimension Style dialog box.

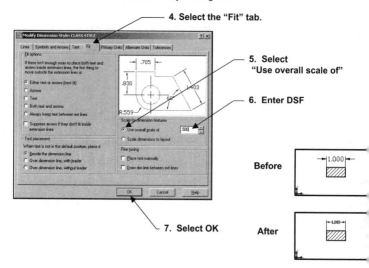

4. Select the "Fit" tab.

5. Select "Use overall scale of"

6. Enter DSF

7. Select OK

Before

After

OW THE DSF EFFECTS HATCH

When you adjust the scale within a Viewport the spacing between the Hatch lines will appear either larger or smaller depending on the scale.

Normal	Too Small	Hatch scale adjusted
1 : 1	2 : 1	2 : 1

REDEFINED

Calculate the Drawing Scale Factor and enter the DSF in the **Scale** box.

Enter DSF here.

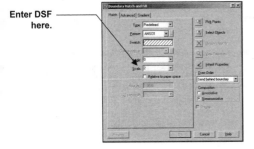

USER DEFINED

Calculate the Drawing Scale Factor and enter the DSF in the **SPACING** box.

Enter DSF here

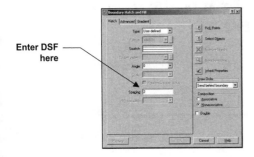

HOW THE DSF EFFECTS TEXTS

When you adjust the scale within a Viewport, any text within the Viewport (in model space) will appear either larger or smaller, depending on the adjusted scale.

To calculate the correct text height, do the following:
1. Decide what height you would like the text to be when plotted on the paper.
2. Calculate the Drawing Scale Factor as follows:
   Multiply the "Drawing Scale Factor" times the "Plotted Text height".

For example:
If you want the text height after plotting to be **1/8"**, and the
adjusted scale within the viewport is 4 : 1 (<u>DSF calculated = **1/4**</u>)
Multiply 1/8 (Plotted Text Height) X 1/4 (DSF) = **1/32"**

So the text height you use is 1/32". Now I know that seems like it will be very small, but remember, the viewport scale has been adjusted to 4 : 1 (4X it's original appearance. So if you set a text height of 1/32" it will be enlarged 4X and have an appearance of 1/8". (That is the height you want when it is plotted)

The following is an example of how the text will appear when plotted.

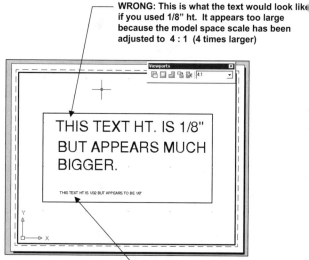

WRONG: This is what the text would look like if you used 1/8" ht. It appears too large because the model space scale has been adjusted to 4 : 1 (4 times larger)

CORRECT: This text appears correct because you took into consideration the adjusted scale.
The text is 1/32 ht times 4 = 1/8

ASSIGN LINEWEIGHTS TO COLORS
Version 2007/2006 shown (Version 2005 and 2004 very similar)

Lineweights may also be assigned to colors within the "color dependent plot style table". Some AutoCAD users prefer to assign lineweights to colors rather than layers or individual objects. As you get more familiar with Lineweights and plotting, you can determine which process you prefer.

To assign lineweights to colors, you must create a "Color dependent Plot Style Table". When this plot style table is selected, it will override the lineweights assigned within the drawing. The step by step process is described below.

1. Select **FILE / PLOT STYLE MANAGER**

2. Select **"Add-A-Plot Style Table" Wizard**

The following dialog boxes will appear.

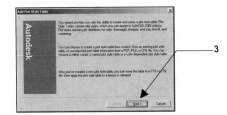

3. Select the **Next** button.

4. Select **"Start from Scratch"** then the **Next** button.

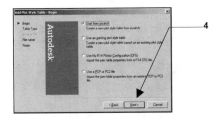

5. Select **"Color-Dependent Plot Style Table"** then the **Next** button.

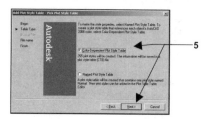

6. Type the new Plot Style Table **name** then select the **Next** button.

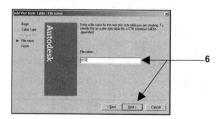

7. Select the **"Plot Style Table Editor"** button.

8. Make changes to the **"PROPERTIES"** then select the **Save & Close** button.

8a. Select color 1

8b. Change Lineweight for the color selected.

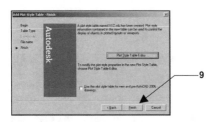

8c

9. Select the **FINISH** button.

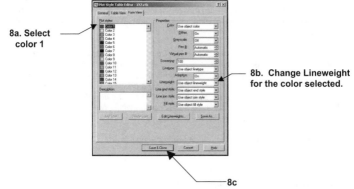

9

10. Select **File / Plot Style Manager**.

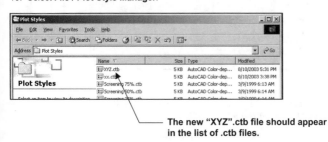

The new **"XYZ"**.ctb file should appear in the list of .ctb files.

HOW TO SAVE A DRAWING

After you have completed a drawing, it is very important to save it. Learning how to save a drawing correctly is almost more important than making the drawing. If you can't save correctly, you will lose the drawing and hours of work.

*There are 2 commands for saving a drawing: **Save** and **Save As**. I prefer to use **Save As**. The **Save As** command always pauses to allow you to choose where you want to store the file and what name to assign to the file. This may seem like a small thing, but it has saved me many times from saving a drawing on top of another drawing by mistake. The **Save** command will automatically save the file either back to where you retrieved it or where you last saved a previous drawing. Neither may be the correct destination. So play it safe, use **Save As** for now.*

1. Start the command by using one of the following methods:

TYPING:	**SAVEAS <enter>**
PULLDOWN:	**FILE / SAVEAS**
TOOLBAR:	**No icon for Save As, only for Save**

This Dialog box should appear:

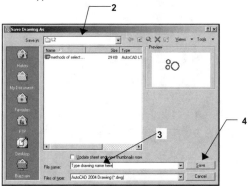

2. Select the appropriate drive and directory from the **"SAVE IN"** box.
    (This is where your drawing will be saved)

3. Type the new drawing file name in the **"FILE NAME"** box.

4. Select the **"SAVE"** button.

SELECTING OBJECTS

Most AutoCAD commands prompt you to "select objects". This means select the objects that you want the command to effect.

There are 2 methods. **Method 1. Pick**, is very easy and should be used if you have only 1 or 2 objects to select. **Method 2. Window**, is a little more difficult but once mastered it is extremely helpful and time saving. Practice the examples shown below.

Pick box

Method 1. PICK :
When the command line prompt reads, "Select Objects", place the cursor (pick box) on top of the object and click the left mouse button. The object about to be selected will highlight.
This appearance change is called "Rollover Highlighting".
This gives you a selection preview of which object will be selected.

Method 2. WINDOW: Crossing and Window

Crossing:
Place your cursor in the area <u>up</u> and to the <u>right</u> of the objects that you wish to select (**P1**) and press the left mouse button. Then move the cursor down and to the left of the objects (**P2**) and press the left mouse button again.
(Note: The window will be *green* and outer line is *dashed*.)
Only the objects that this window **crosses or completely enclosed** will be selected.

In the example on the right, all 3 circles have been selected. (The 2 small circles are **completely enclosed** and the large circle is **crossed** by the window.)

Window:
Place your cursor in the area <u>up</u> and to the <u>left</u> of the objects that you wish to select (**P1**) and press the left mouse button. Then move the cursor <u>down</u> and to the <u>right</u> of the objects (**P2**) and press the left mouse button.
(Note: The window will be *blue* and outer line is *solid*.)
Only the objects that this window **completely enclosed** will be selected.

In the example on the right, only 2 circles have been selected. (The large circle is **not** completely enclosed.)

Note: If the windows described above do not show up on your screen, it means that your "implied windowing" is turned off. Select <u>Tools / Options / Selection tab</u>. In the section "<u>Selection Modes</u>" on the left, place a check mark in the "<u>Implied windowing</u>" box. Also, the colors and the opacity of the windows may be customized using the "Visual Effect Settings".

STARTING A NEW DRAWING

1. Start the command using one of the following methods:

TYPING: **NEW <enter> or press CTRL + N**
PULLDOWN: **FILE / NEW**
TOOLBAR: **STANDARD**

The Dialog box shown below should appear.

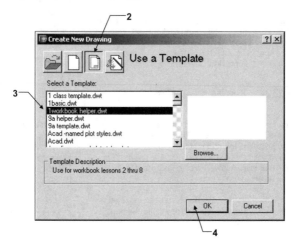

2. Select the **Use a Template** box (third from the left).

3. Select a template from the list of templates.

4. Select the **OK** button (bottom right).

TABLES – 2007 & 2006 only

A Table is an object that contains data organized within columns and rows. AutoCAD's Table feature allows you to modify an existing Table Style or create your own Table Style and then enter text or even a block into the table cells. This is a very simple to use feature with many options.

This is an example of a Table.

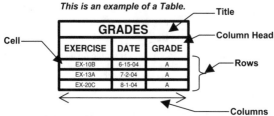

HOW TO CREATE A TABLE.

1. Select the Table Style command using one of the following:

   TYPE = tablestyle
   PULLDOWN = FORMAT / TABLE STYLE
   TOOLBAR = FORMAT

The following dialog box will appear.

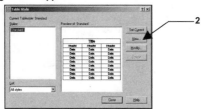

2. Select the **NEW** button. *The following dialog will appear.*

3. Enter the new Table Style name. *Note: When you create a new table style you always "Start With" an existing style and you specify the differences.*

4. Select the **Continue** button.

5. Specify table properties for the Title, Column Heads and the Data.
    Notice the **3 tabs** in the upper left corner. The properties within each of these tabs is the same except for the following:

The <u>check mark</u> indicates that you want a "**Header for the Column**" and/or a "**Title Header**". If you do not want one or both, remove the check mark. The setting boxes on that tab will turn gray and you will not be required to enter information for that tab.

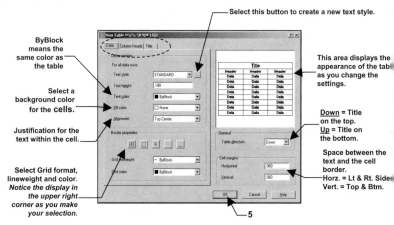

Select this button to create a new text style.

ByBlock means the same color as the table

Select a background color for the **cells.**

Justification for the text within the cell.

Select Grid format, lineweight and color. *Notice the display in the upper right corner as you make your selection.*

This area displays the appearance of the tab as you change the settings.

<u>Down</u> = Title on the top.
<u>Up</u> = Title on the bottom.

Space between the text and the cell border.
Horz. = Lt & Rt. Sides
Vert. = Top & Btm.

5. Select the **OK button** after you have completed all the settings necessary shown above.

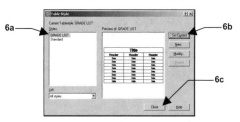

6. Select the **(a) New Table Style** and select the **(b) Set Current**" button and **(c) close.**

OW TO INSERT A TABLE

Select **DRAW / TABLE.**

he following dialog will appear.

Click this button to create a new Table Style.

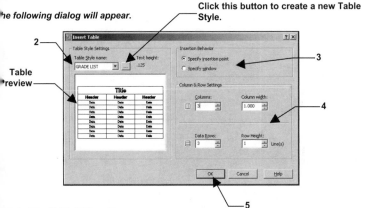

Select the **Table Style name**.

Select the Insertion Behavior.

Specify Insertion Point: Specify the Columns, Column width, Rows and Rows Height. When you select the OK button you select the location to insert the table. The size of the table is determined by the settings you specified.

Specify window: Specify the number of Columns and Row height. When you select the OK button you select the location for the upper left corner of the table. Then drag the cursor to specify the Column width and number of Rows, on the screen.

Specify the Column and Row specifications.

Select **OK.**

Place the insertion point. (Refer to #3 above.)

The Table is now on the screen waiting for you to fill in the data. Use the Tab or Arrow keys to move between the cells.

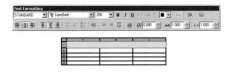

When you have filled all of the cells, select **OK.**

HOW TO INSERT A BLOCK INTO A TABLE CELL

1. Left click in the cell you wish to insert a block.

2. Select Insert Block from the menu.

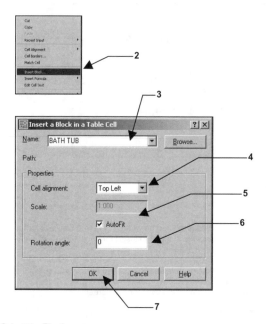

3. Select the **Block name**.

4. Select the **Cell alignment**.

5. Select the **Scale.**
   Note: <u>AutoFit</u> will automatically size the block to fit within the cell.

6. Select **Rotation angle**.

7. Select the **OK** button.

may apply simple numerical operations such as Sum, Average, Count, set cells equal to
r cells or even add an equation of your own.

following examples are for Sum and Average operations.

M

lick in the Cell that you wish to enter a formula.

ROOM	TABLES	CHAIRS	COST
B14	2	4	100
F22	3	6	400
G7	4	8	500
TOTAL			
AVERAGE			

———1

Right click and select:
Insert Formula / Sum

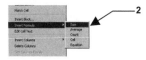

———2

Select the cells content to sum using window.

ROOM	TABLES	CHAIRS	COST
B14	2	4	100
F22	3	6	400
G7	4	8	500
TOTAL			
AVERAGE			

———3

Verify the formula and select OK.
You may edit the formula if necessary.

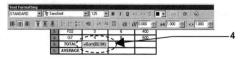

———4

ce the formula disappears and the sum of the cells selected has been calculated. Also the
e is shaded to make you aware that this cell has a formula in it.

ROOM	TABLES	CHAIRS	COST
B14	2	4	100
F22	3	6	400
G7		8	500
TOTAL	9		
AVERAGE			

———Sum

AVERAGE

1. Set the precision in **Format / Units**.
   Note: It is important to set the precision before you start the operation. This is only necessary for the Average operation. Each cell that you assign the Average formula to may have a different precision.

2. Click in the Cell that you wish to enter a formula.

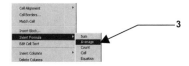

3. Right click and select:
   Insert Formula / Average

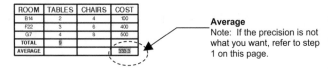

4. Select the cells content to average using a window.

ROOM	TABLES	CHAIRS	COST
B14	2	4	100
F22	3	6	400
G7	4	8	500
TOTAL	9		
AVERAGE			

5. Verify the formula and select OK.
   You may edit the formula if necessary.

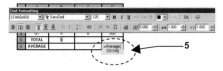

Notice the formula disappears and the average of the cells selected has been calculated.

ROOM	TABLES	CHAIRS	COST
B14	2	4	100
F22	3	6	400
G7	4	8	500
TOTAL	9		
AVERAGE			333.3

Average
Note: If the precision is not what you want, refer to step 1 on this page.

W TO MODIFY AN EXISTING TABLE

T TEXT WITHIN A CELL.
)ouble click inside the cell. (Be careful not to click on a border line.)
The text formatting dialog box will appear.

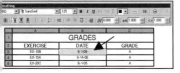

2. Make the text changes.
3. Select the OK button.

ANGE THE COLUMN WIDTH OR ROW HEIGHT.
Click once inside the cell or row that you wish to change.
Right click and select "Properties" from the menu.

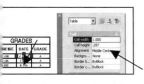

3. In the Properties Palette under Cell, click the cell width or the cell height value and enter the new value.

D A COLUMN OR ROW.
Click once inside a cell where you wish to add a column or a row.
Right click and select one of the following options from the menu:

COLUMNS:
Right: Inserts a column to the right of the selected cell.
Left: Inserts a column to the left of the selected cell.

ROWS:
Above: Inserts a row above the selected cell.
Below: Inserts a row below the selected cell.

LETE A COLUMN OR ROW.
Click once inside one of the cells in the column or row that you wish to delete.
Right click and select Delete Columns or Delete Rows from the menu.

MODIFY A TABLE USING GRIPS

You may also modify tables using Grips. When editing with Grips, the left edge of the table remains stationary but the right edge can move. The upper left Grip is the Base Point for the table.

To use Grips, click on a table border line. The Grips should appear. Each Grip has a specific duty, shown below. To use a Grip, click on the Grip and it will change to red. Now click and drag it to the desired location.

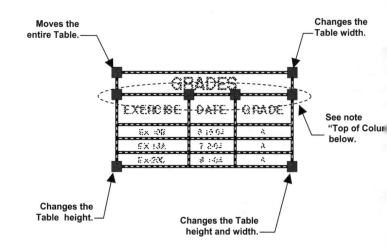

Moves the entire Table.

Changes the Table width.

See note "Top of Colum below.

Changes the Table height.

Changes the Table height and width.

Top of Column
There is a Grip located at the top of each column line. These Grips adjust the width of the column to the left of the Grip. The entire Table adjusts at the same time. If you hold the CTRL key down while moving a column Grip, column will change but the width of the entire Table remains unchanged.

LDS – 2007 & 2006 only   (Not available in LT version)

Field is a string of text that has been set up to display data that it gets from another source. For example, you may create a field that will display the Circumference of a specific Circle in your drawing. If you changed the diameter of that Circle you could "update" the field and it would display the new Circumference.

Fields can be used in many different ways. After you have read through the example below you will understand the steps required to create and update a Field. Then you should experiment with some of the other Field Categories to see if they would be useful to you. Consider adding Fields to a cell within a Table.

CREATE A FIELD

Draw a 2" Diameter Circle and place it anywhere in the drawing area.
Select "**Insert / Field...**".

3. Select the Field
   Category: Objects

Select the Field Name:
   Object

5. Select the Object
   Type button.

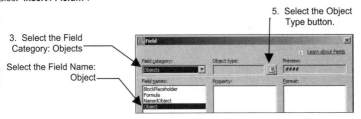

Select the Circle that you just drew.

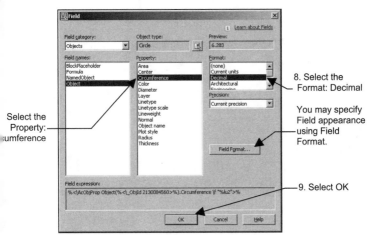

Select the
Property:
Circumference

8. Select the
   Format: Decimal

You may specify
Field appearance
using Field
Format.

9. Select OK

5-41

10. Place the "Field" inside the Circle.

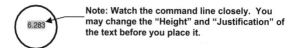

Note: Watch the command line closely. You may change the "Height" and "Justification" of the text before you place it.

Notice that the Field appears with a gray background. This background will not plot. The background can be turned off but I find it helpful to be able to visually distinguish a Field from plain text. If you wish to turn it off use: Tools / Options / User Preferences tab. In the Fields section, uncheck the "Display background of Fields" box.

UPDATE A FIELD

Now let's change the diameter of that Circle and see what happens to the Field.

1. Change the size of the Circle.
   a. Select the Circle.
   b. Right click and select **Properties.**
   c. Change the Diameter to 4.
   d. Close the Properties Palette.
   e. Press ESC key to clear the Grips.

Notice that the Field has not changed yet.
In order to see the new Circumference value, you must "Update the Field".

2. Select **Tools / Update Fields** or **View / Regen.**

3. Select the Field and <enter>.

12.566

The Field updated to the new Circumference value.

Note: The Field will update automatically each time you Save, Plot or Regen the drawing.

iting FIELDS

ng a Field is very easy. The process is basically the same as creating a Field.

T A FIELD

ouble click the **Field text**. *The Multiline Text Editor will open.*
ght click on the **Field text**.
elect **Edit Field** from the menu.
ake the changes.
elect **OK**.
ou may also add more text to the Field text but you may not change the text within the
eld.
xample: Add the words "Circle Circumference" under the Field text on the previous pages.
elect **OK** to exit the Multiline Text Editor.

) A FIELD TO A TABLE CELL

elect the cell.
ght click.
elect "Insert Field" from the menu.
reate a Field.

---

estions about Fields

/hat happens when you Explode a Field?

he Field will convert to normal text and will no longer update.

/hat happens if someone opens the drawing in another version like AutoCAD 2000 r LT?

he last value displayed will appear but it will not be a Field.

/ill all of my Fields be ruined now?

lo. When you open the drawing in AutoCAD 2006 all the Fields should return.

Section 6
Layers

LAYERS

A **LAYER** is like a transparency. Have you ever used an overhead light projector? Remember those transparencies that are laid on top of the light projector? You could stack multiple sheets but the projected image would have the appearance of one document. Layers are basically the same. Multiple layers can be used within one drawing.

The example, on the right, shows 3 layers. One for annotations (text), one for dimensions and one for objects.

It is good "drawing management" to draw related objects on the same layer. For example, in an architectural drawing, you could have the walls of a floor plan on one layer and the Electrical and Plumbing on two other layers. These layers can then be Thawed (ON) or Frozen (OFF) independently. If a layer is Frozen, it is not visible. When you Thaw the layer it becomes visible again. This will allow you to view or make plots with specific layers visible or invisible.

SELECTING A LAYER - Method 1. (Also refer to Method 2 on the next page)

1. Display the LAYER CONTROL DROP-DOWN LIST below by clicking on the down arrow. (▼)

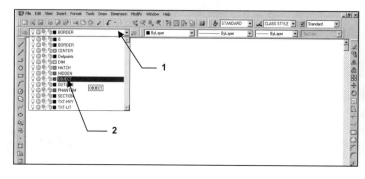

2. Click on the Layer Name you wish to select. The Layer selected will become the Current layer and the drop-down list will disappear. The Current layer means that the next object drawn will reside on this layer and will have the same color, linetype and lineweight. These are called Properties.

SELECTING A LAYER - Method 2 for Version 2007, 2006 and 2005.

1. Select the Layer command using one of the following:

   TYPE = LA <enter>
   PULLDOWN = FORMAT / LAYER
   TOOLBAR = OBJECT PROPERTIES

2. The "Layer Properties Manager" dialog box, shown below, will appear.
3. First select a layer by Clicking on its name.
4. Select the **CURRENT** button. (The green check mark)
5. Then select the **OK** button.

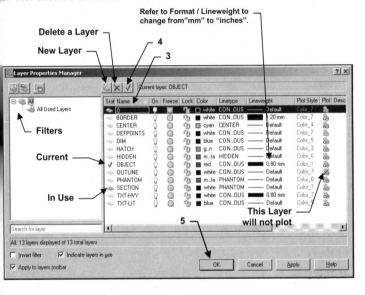

The layer you have just selected is now the **CURRENT** layer. This means that the next object drawn will reside on this layer and will have the same color, linetype and lineweight. These are called <u>Properties.</u>

How to delete a layer: Left click on the layer name, right click, select "Delete Layer" from the options menu. <u>Note: You can't delete the "current" layer or a layer in use.</u>

6-3

SELECTING A LAYER - Method 2 for Version <u>2004</u>.

1. Select the Layer command using one of the following:

   TYPE = LA
   PULLDOWN = FORMAT / LAYER
   TOOLBAR = OBJECT PROPERTIES

2. The dialogue box below will appear.

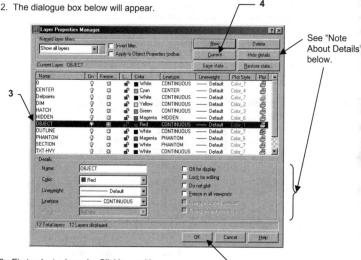

See "Note About Details" below.

3. First select a layer by Clicking on it's name.
4. Select the **CURRENT** button.
5. Then select the **OK** button.

The layer you have just selected is now the **CURRENT** layer. This means that the next object drawn will reside on this layer and will have the same properties as this layer. Notice that the layers shown above have specific colors, linetypes and lineweights. These are called Properties.

Note about "Details":

If your dialog box does not have the "**Details**" area at the bottom, select the "Show details" button. When you select the "Show details" button, it will change to a "Hide details" button, as shown above.

CREATING NEW LAYERS – Version 2004 only

1. Select the Layer command using one of the following:

> **TYPE = LA**
> **PULLDOWN = FORMAT / LAYER**
> **TOOLBAR = OBJECT PROPERTIES**

The Layer & Linetype Properties dialog box shown below will appear.

2. Select the NEW button. A new layer will appear named "Layer1"

Note: If the "Details" area is not shown on your dialog box, select the "Show Details" box.

Layer 1 →

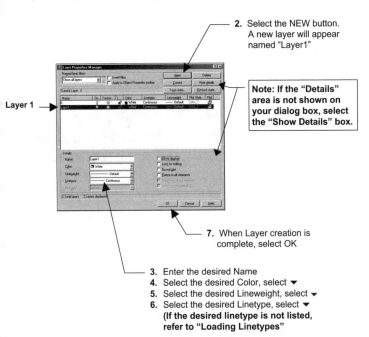

7. When Layer creation is complete, select OK

3. Enter the desired Name
4. Select the desired Color, select ▼
5. Select the desired Lineweight, select ▼
6. Select the desired Linetype, select ▼
(If the desired linetype is not listed, refer to "Loading Linetypes"

CREATING NEW LAYERS- Version 2007, 2006 and 2005

Using layers is an important part of managing and controlling your drawing. It is better to have too many layers than too few. Draw like objects on the same layer.
For example, place all doors on the layer "door" or centerlines on the layer "centerline". When you create a new layer you will assign a name, color, linetype, lineweight and whether or not it should plot.

1. Select the Layer command using one of the following:

> **TYPE = LA**
> **PULLDOWN = FORMAT / LAYER**
> **TOOLBAR = OBJECT PROPERTIES**

The Layer Properties Manager dialog box shown below will appear.

2. NEW LAYER button

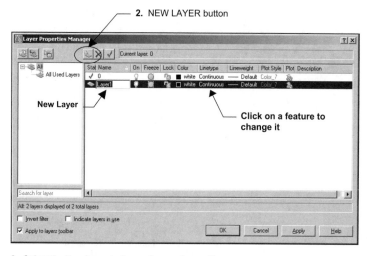

2. Select the New Layer button and a new layer will appear.
   Type the new layer name and press <enter>

3. Click on any of the features and a dialog box will appear.

The following pages will describe <u>Color</u>, <u>Linetype</u>, and <u>Layer Controls</u>.

COLOR – Version 2007, 2006 and 2005

Color is not merely to make a pretty display on the monitor screen.

Here are some things to consider when selecting the colors for your layers.

1. Now that color printers are so commonplace, consider how the colors will appear on the paper. (Pastels do not appear well on white paper.)

2. Consider how the colors will appear on the screen. (Yellow appears well on a black background but not on white. Changing the screen color is in the Advanced Workbook.)

3. If you choose to use color dependent plot styles instead of lineweights, color will determine the width of the lines.

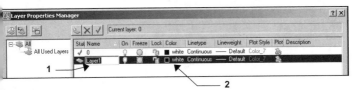

1. Select (highlight) the layer that you want to change.

2. Select the color swatch or word. (white)

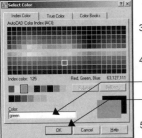

3. Select the color (The name or the number will appear in the color box.)

4. Select the **OK** button.

5. The color selected will appear on the layer line.

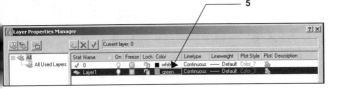

LOADING A LINETYPE – Version 2004 only

In an effort to conserve data within a drawing file, AutoCAD automatically loads only one linetype called "continuous". If you would like to use other linetypes, such as "dashed", you must "Load" them into the drawing as follows.

1. Select the **LINETYPE** command using one of the following:

> **TYPE = LT**
> **PULLDOWN = FORMAT / LINETYPE**
> **TOOLBAR = NONE**

The Linetype Manager dialog box shown below will appear.

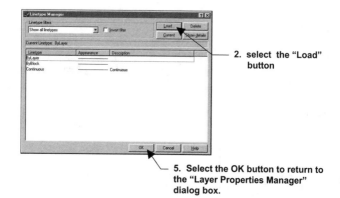

2. select the "Load" button

5. Select the OK button to return to the "Layer Properties Manager" dialog box.

*The **Load or Reload Linetypes** dialogue box shown below will appear.*

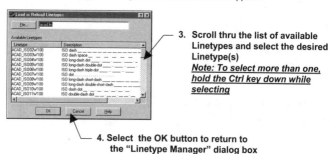

3. Scroll thru the list of available Linetypes and select the desired Linetype(s)
Note: To select more than one, hold the Ctrl key down while selecting

4. Select the OK button to return to the "Linetype Manager" dialog box

LOADING and SELECTING a LINETYPE – 2007, 2006 & 2005

In an effort to conserve data within a drawing file, AutoCAD automatically loads only one linetype called "continuous". If you would like to use other linetypes, such as "dashed", you must "Load" them into the drawing as follows.

1. Select the Linetype.

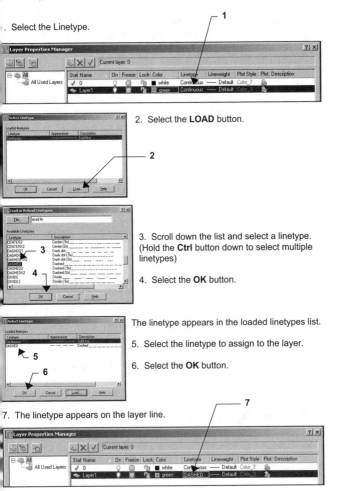

2. Select the **LOAD** button.

3. Scroll down the list and select a linetype. (Hold the **Ctrl** button down to select multiple linetypes)

4. Select the **OK** button.

The linetype appears in the loaded linetypes list.

5. Select the linetype to assign to the layer.

6. Select the **OK** button.

7. The linetype appears on the layer line.

LAYER CONTROL DEFINITIONS – Version 2004 only

Off For Display

If a layer is **ON** it is **visible**. If a layer is **OFF** it is **not visible**.
Only layers that are **ON** can be **edited** or **plotted**.
(Warning: Objects on a Layer that is OFF can be <u>accidentally erased</u> even though they
are invisible. When you are asked to select objects, in the erase command, if you type
ALL <enter> all objects will be selected even the invisible ones.)

Lock For Editing

LOCKED layers are visible but <u>cannot be edited</u>. They are visible so they **will** be
plotted. (Locked layers <u>cannot be selected</u> by typing ALL.)

Do Not Plot

This option prevents a layer from plotting even though it is visible.

Freeze In All Viewports

Freeze and **Thaw** are very similar to On and Off. A Frozen layer is <u>not visible</u>.
A Thawed layer <u>is visible</u>. Only thawed layers can be edited or plotted.

Additionally:
a. Objects on a Frozen layer **cannot** be accidentally erased by typing All.
b. Freezing saves time, when working with large and complex drawings, because
   frozen layers are not **regenerated** when you zoom in and out.

Freeze In Current Viewport

Layers selected will be frozen in the "current" viewport only. Current means the active
viewport. Only one viewport can be active at one time.
(Available in Paperspace only)

Freeze In New Viewports

Layers selected will be frozen in the next viewport created. So you are selecting the
layers, to be frozen, before you have created the viewport. This one doesn't get used
much.
(Available in Paperspace only)

LAYER CONTROL DEFINITIONS – 2007, 2006 & 2005

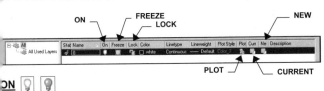

ON

If a layer is **ON** it is **visible**. If a layer is **OFF** it is **not visible**.
Only layers that are **ON** can be **edited** or **plotted**.
Warning: Objects on a Layer that is OFF can be <u>accidentally erased</u> even though they
are invisible. When you are asked to select objects in the erase command, if you type
ALL <enter> all objects will be selected; even the invisible ones.)

LOCK

LOCKED layers are visible but <u>cannot be edited</u>. They are visible so they **will** be
plotted. (Locked layers <u>cannot be selected</u> by typing ALL.)

PLOT

This option prevents a layer from plotting even though it is visible.

FREEZE

Freeze and **Thaw** are very similar to On and Off. A Frozen layer is <u>not visible</u>.
A Thawed layer <u>is visible</u>. Only thawed layers can be edited or plotted.

Additionally:
a. Objects on a Frozen layer **cannot** be accidentally erased by typing All.
b. When working with large and complex drawings, freezing saves time because frozen
   layers are not **regenerated** when you zoom in and out.

Note: The next 2 are displayed only when you select a Layout tab.

CURRENT

Layers selected will be frozen in the "current" viewport only. Current means the active
viewport. Only one viewport can be active at one time.
(Available in Paper Space only)

NEW

Layers selected will be frozen in the next viewport created. So you are selecting the
layers to be frozen before you have created the viewport. This one doesn't get used
much.
(Available in Paper Space only)

Section 7
Input Options

COORDINATE INPUT

Autocad uses the *Cartesian Coordinate System.*

The Cartesian Coordinate System has 3 axes, X, Y and Z.

The **X** is the Horizontal axis. (*Right and Left*)
The **Y** is the Vertical axis. (*Up and Down*)
The **Z** is Perpendicular to the X and Y plane.
(*The Z axis, <u>which is not discussed in this book</u>, is used for 3D.*)

UCS icon

Look at the User Coordinate System (UCS) icon in the lower left corner of your screen.
The arrows are pointing in the positive direction.

The location where the X , Y and Z axes intersect is called the **ORIGIN**. (0,0,0)
Currently the Origin is located in the lower left corner of the screen.
When you move the cursor away from the Origin, in the direction of the arrows, the X
and Y coordinates are positive.
When you move the cursor in the opposite direction, the X and Y coordinates are
negative.

Using this system, every point on the
screen can be specified using positive
and negative X and Y coordinates.

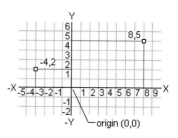

FOR VERSION 2007 & 2006 only
Note: Please confirm that
Dynamic Input is "Off" by releasing
the "DYN" button in the status bar
or press F12.
Dynamic Input will be discussed in
Lesson 11.

There are <u>3 types of Coordinate input</u>, **Absolute, Relative** and **Polar**.
(Polar will be discussed in Lesson 11)

ABSOLUTE COORDINATES   (The input format is: **X, Y**)

Absolute coordinates come *from the ORIGIN* and are typed as follows: **8, 5** .
The first number (8) represents the **X-axis** (horizontal) distance <u>from the Origin</u> and the
second number (5) represents the **Y-axis** (vertical) distance from the Origin.
The two numbers must be separated by a **comma**.

An absolute coordinate of **4, 2** will be **4** units to the right (horizontal) and **2** units up
(vertical) <u>from the current location of the Origin</u>.

An absolute coordinate of **-4, -2** will be **4** units to the left (horizontal) and **2** units down
(vertical) <u>from the current location of the Origin</u>.

The following are examples of Absolute Coordinate input.
<u>Notice where the Origin is located in each example.</u>

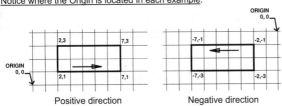

Positive direction   Negative direction

RELATIVE COORDINATES (The input format is: **@X, Y**)

Relative coordinates come *from the last point entered*. The first number represents
the **X-axis** (horizontal) and the second number represents the **Y-axis** (vertical).
The two numbers must be separated by a **comma**. To distinguish between Absolute
and Relative, use the **@** symbol and then the X and Y coordinates.

A Relative coordinate of **@5, 2** will go to the **right** 5 units and **up** 2 units
<u>from the last point entered</u>.

A Relative coordinate of **@-5, -2** will go to the **left** 5 units and **down** 2 units
<u>from the last point entered</u>.

The following is an example of Relative Coordinate input.

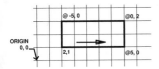

EXAMPLES OF COORDINATE INPUT

Scenario 1.
You want to draw a line with the first endpoint "at the Origin" and the second endpoint 3 units in the positive X direction.

1. Select the Line command.
2. You are prompted for the first endpoint: **Type 0, 0 <enter>**
3. You are then prompted for the second endpoint: **Type 3, 0 <enter>**

What did we do?
The first endpoint coordinate input, 0,0 means that you do not want to move away from the Origin. You want to start "on" the Origin.

The second endpoint coordinate input, 3, 0 means that you want to move 3 units in the positive X axis. The "0" means you do not want to move in the Y axis. So the line will be exactly horizontal.

Scenario 2.
You want to start a line 8 units directly above the origin and it will be 4 units in length, perfectly vertical.

1. Select the Line command.
2. You are prompted for the first endpoint: **Type 0, 8 <enter>**
3. You are prompted for the second endpoint: **Type @0, 4 <enter>**

What did we do?
The first endpoint coordinate input, 0, 8 means you do not want to move in the X axis direction but you do want to move in the Y axis direction.

The second endpoint coordinate input @0, 4 means you do not want to move in the X axis "from the last point entered" but you do want to move in the Y axis "from the last point entered. (Remember the @ symbol tells AutoCAD that the coordinates typed are relative to the "last point entered" not the Origin.

Scenario 3.
Now try drawing 5 connecting lines segments.

1. Select the Line command.
2. First endpoint: 2, 4 <enter>
3. Second endpoint: @ 2, -3 <enter>
4. Second endpoint: @ 0, -1 <enter>
5. Second endpoint: @ -1, 0 <enter>
6. Second endpoint: @ -2, 2 <enter>
7. Second endpoint: @ 0, 2 <enter> <enter>

Notice the "@" symbol for relative coordinates.

Note: If you enter an incorrect coordinate, just type "U" <enter> and the last segment will disappear and you will have another chance at entering the correct coordinate.

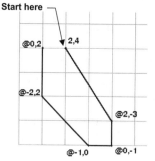

Start here

@0,2   2,4

@-2,2

@2,-3

@-1,0   @0,-1

DIRECT DISTANCE ENTRY (DDE)

DIRECT **D**ISTANCE **E**NTRY is a combination of keyboard entry and cursor movement. **DDE** is used to specify distances in the horizontal or vertical axes <u>from the last point entered</u>. **DDE** is a **_Relative Input_**. Since it is used for Horizontal and Vertical movements, **Ortho** must be **ON**.

Using DDE is simple. Just move the cursor and type the distance.
Negative and positive is understood automatically by moving the cursor up (positive), down (negative), right (positive) or left (negative) from the last point entered. No minus sign necessary.

Moving the cursor to the right and typing 5 and <enter> tells AutoCAD that the 5 is positive and Horizontal.
Moving the cursor to the left and typing 5 and <enter> tells AutoCAD that the 5 is negative and Horizontal.
Moving the cursor up and typing 5 and <enter> tells AutoCAD that the 5 is positive and Vertical.
Moving the cursor down and typing 5 and <enter> tells AutoCAD that the 5 is negative and Vertical.

EXAMPLE:

1. <u>Ortho must be ON</u>.
2. Select the Line command.
3. Type: 1, 2 <enter> to enter the first endpoint using Absolute coordinates.
4. Now move your cursor to the right and type: 5 <enter>
5. Now move your cursor up and type: 4 <enter>
6. Now move your cursor to the left and type: 5 <enter> <enter> to stop

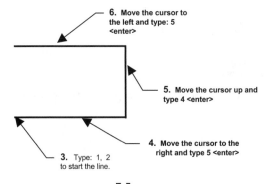

6. Move the cursor to the left and type: 5 <enter>

5. Move the cursor up and type 4 <enter>

4. Move the cursor to the right and type 5 <enter>

3. Type: 1, 2 to start the line.

DYNAMIC INPUT – Version 2007 & 2006 only

To help you keep your focus in the "drawing area", AutoCAD has provided a command interface called **Dynamic Input**.

When Dynamic Input is ON abbreviated prompts that you normally see on the command line are displayed near the cursor. You may also input information within the Dynamic Input instead of on the command line.

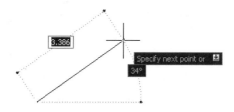

Note: *Some users find Dynamic Input useful some find it distracting. After completing this lesson you decide if you want to use it. It is your choice.*

How to turn Dynamic Input ON or OFF

Select the **DYN** button on the status bar or use the F12 key.

DYNAMIC INPUT has 3 components

1. **Pointer Input**
2. **Dimensional Input**
3. **Dynamic Prompts**

You may control what is displayed by each component and turn each ON or OFF.

POINTER INPUT

Pointer Input is only displayed for the **first** point.
When Pointer Input is enabled (ON) and a command has been selected, the location of the crosshairs is displayed as coordinates in a tooltip near the cursor.

Example:

1. Select the **LINE** command.

2. Move the cursor.

The 2 boxes display the cursor location as X and Y coordinates **from the Origin**. (Absolute coordinates)

3. You may move the cursor until these coordinates display the desired location and press the left mouse button or enter the desired coordinate values in the tooltip boxes instead of on the command line.

Note:
Refer to "How to enter coordinate values in Dynamic Input tooltips".

How to change POINTER INPUT settings

1. Right-click the **Dyn** button on the status bar and select **Settings**.

2. In the Drafting Settings dialog box select the **Dynamic input** tab.

3. Under Pointer Input select the **Settings** button.

4. In the Pointer Input Settings dialog box select:

   Format: (select one as the display format)

   Polar or Cartesian format

   Relative or Absolute coordinate

   Visibility: (select one for tooltip display)

   - **As Soon As I Type Coordinate Data**. When pointer input is turned on, displays tooltips only when you start to enter coordinate data.

   - **When a Command Asks for a Point**. When pointer input is turned on, displays tooltips whenever a command prompts for a point.

   - **Always—Even When Not in a Command**. Always displays tooltips when pointer input is turned on.

5. Select **OK** to close each dialog box.

DIMENSIONAL INPUT

When Dimensional Input is enabled (ON) the tooltips display the distance and angle values for the **second** and **subsequent points**.

The default display is Relative Polar coordinates.

Dimensional Input is available for: **Arc, Circle, Ellipse, Line** and **Pline**.

Example:
1. Select the Line Command.

2. Enter the 1st point.

3. You may enter the **distance** and then the **angle** in the tooltip boxes instead of on the command line.

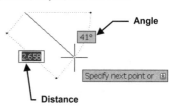

Angle

41°

2.658

Specify next point or

Distance

Note:
Refer to "How to enter coordinate values in Dynamic Input tooltips".

How to change DIMENSIONAL INPUT settings

1. Right-click the **Dyn** button on the status bar and select **Settings**.

2. In the Drafting Settings dialog box select the **Dynamic input** tab.

3. Under Dimension Input, select the **Settings** button.

4. In the Dimension Input Settings dialog box select:

   Visibility: (select one of the following options)

   o **Show Only 1 Dimension Input Field at a Time**. Displays only the distance dimensional input tooltip when you are using grip editing to stretch an object.

   o **Show 2 Dimension Input Fields at a Time**. Displays the distance and angle dimensional input tooltips when you are using grip editing to stretch an object.

   o **Show the Following Dimension Input Fields Simultaneously**. Displays the selected dimensional input tooltips when you are using grip editing to stretch an object. Select one or more of the check boxes.

5. Select **OK** to close each dialog box.

DYNAMIC PROMPTS

When Dynamic Prompts are enabled (ON) prompts are displayed.

You may enter a response in the tooltip box instead of on the command line.

Example:
1. Select the <u>Circle</u> command.

2. Place the <u>center point</u> for the circle
   by entering the X and Y
   coordinate values in the tooltip
   boxes or press the left mouse
   button.

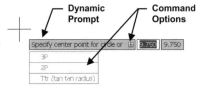

3. Press the right mouse button for command options or you may also press the down
   arrow to view command options.
   Select an option by clicking on it or press the up arrow for option menu to disappear.

4. Enter Radius or select Diameter using one of the methods in 3 above.

How to change the COLOR, SIZE, or TRANSPARENCY of tooltips

1. Right-click the **Dyn** button on the status bar then select **Settings**.

2. In the Drafting Settings dialog box select the **Dynamic input** tab.

3. At the bottom of the dialog box select **Drafting Tooltip Appearance** button.

4. Under Color, select the Model Color or Layout Color button to display the **Select Color** dialog box. You may specify a color for tooltips in the space you selected.

5. Under **Size**, move the slider to the right to make tooltips larger or to the left to make them smaller. The default value, 0, is in the middle.

6. Under **Transparency**, move the slider. The higher the setting, the more transparent the tooltip.

7. Under **Apply To**, choose an option:

   o **Override OS Settings for All Drafting Tooltips**. Applies the settings to all tooltips, overriding the settings in the operating system.

   o **Use Settings Only for Dynamic Input Tooltips**. Applies the settings only to the drafting tooltips used in Dynamic Input.

8. Select **OK** to close each dialog box.

HOW TO ENTER COORDINATE VALUES IN DYNAMIC INPUT TOOLTIP BOXES

Use one of the following methods.

To enter Cartesian coordinates (X and Y)

1. Enter an "**X**" coordinate value <u>and a **comma**</u>.

2. Enter an "**Y**" coordinate value <enter>.

To enter Polar coordinates (from the last point entered)

1. Enter the **distance** value from the last point entered.

2. Press the **Tab** key.

3. Move the cursor in the approximate direction and enter the **angle** value <enter>

Note: Enter an angle value of <u>0-180</u> only and the angle is always <u>positive</u>.

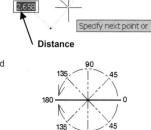

How to specify Absolute or Relative coordinates

<u>To enter absolute coordinates</u> when relative coordinate format is displayed in the tooltip. Enter **#** to temporarily override the setting. (shift key + 3)

<u>To enter relative coordinates</u> when absolute coordinate format is displayed in the tooltip. Enter **@** to temporarily override the setting. (shift key + 2)

<u>To enter absolute world coordinates</u> (WCS), enter **✱**

Note about Ortho
You may toggle Ortho On and OFF by holding down the **shift** key.
This is a easy method to use Direct Distance Entry while using Dynamic Input.

POLAR COORDINATE INPUT

UNDERSTANDING THE *"POLAR DEGREE CLOCK"*

The Angle of the line determines the direction. For example: If you want to draw a line
at a 45 degree angle towards the upper right corner, you would use the angle 45. But if
you want to draw a line at a 45 degree angle towards the lower left corner, you would
use the angle 225. You may also use Polar Input for Horizontal and Vertical lines using
the angles 0, 90, 180 and 270. No negative input is required.

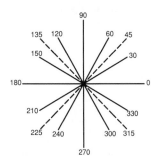

DRAWING WITH *POLAR COORDINATE INPUT*

A Polar coordinate may come *from the last point entered* or *from the Origin,*
depending upon whether you use the @ symbol or not. The first number represents the
Distance and the second number represents the **Angle**. The two numbers are
separated by the **less than (<)** symbol.
The input format is: **distance < angle**

A Polar coordinate of **@6<45** will be 6 units long and at an angle of 45 degrees *from
the last point entered.*

A Polar coordinate of **6<45** will be 6 units and 45 degrees from the *Origin*.

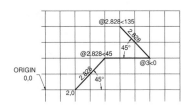

POLAR SNAP

Polar Snap is used with Polar Tracking to make the cursor snap to specific *distances* and *angles*. If you set Polar Snap distance to 1 and Polar Tracking to angle 30 you can draw lines 1, 2, 3, 4 units... long at an angle of 30, 60, 90 etc. without typing anything on the command line. You just move the cursor and watch the tool tips.

SETTING THE ANGLE INCREMENT

1. Right Click on the POLAR button on the Status Bar and select "SETTINGS" or select **Tools / Drafting Settings / Polar Tracking** tab.
   The following dialog box will appear:

2. Select

3. Set the Increment Angle to: 15

LT does not have this option.

4. Select "ABSOLUTE"

5. Select the OK button.

SETTING THE POLAR SNAP

1. Right Click on the SNAP button on the Status Bar and select "SETTINGS" or select **Tools / Drafting Settings / Snap and Grid** tab.
   The following dialog box will appear:

2. Select

4. Set the Polar Spacing distance

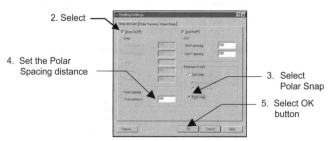

3. Select Polar Snap

5. Select OK button

SNAP AND GRID - Sets standard snap and grid information
POLAR SPACING -Increment Snap distance when Polar Snap is ON.
STYLE - This setting will be taught in the Advanced course.
TYPE - Sets the Snap to Polar or Grid

POLAR TRACKING

Polar Tracking can be used instead of **ORTHO**. When *Polar Tracking* is "**ON**", a dotted *"tracking"* line and a *"tool tip"* box appear. The tracking line.... "snaps" to a **preset angle increment** when the cursor approaches one of the preset angles. The word *"Polar"*, followed by the *"distance"* and *"angle"* from the last point appears in the box.

Tracking line

Tool Tip

SETTING THE ANGLE INCREMENT

1. Right Click on the POLAR button on the Status Bar and select "**SETTINGS**", or select **Tools / Drafting Settings / Polar Tracking** tab.
   The following dialog box will appear:

2. Set the
Increment Angle to: 15.

LT does not have this option.

3. Select "ABSOLUTE".

4. Select the OK button.

POLAR ANGLE SETTINGS

Increment Angle   Choose from a list of Angle increments including 90, 45, 30, 22.5, 18, 15,10 and 5. You will be able to "snap" to multiples of that angle.

Additional Angles   Check this box if you would like to use an angle other than one in the Incremental Angle list. For example: 12.5.

New   You may add an angle by selecting the "New" button. You will be able to snap to this new angle in addition to the incremental Angle selected. But you will not be able to snap to it's multiple. For example, if you selected 7, you would not be able to snap to 14.

Delete   Deletes an Additional Angle. Select the Additional angle to be deleted and then the Delete button.

POLAR ANGLE MEASUREMENT

ABSOLUTE Polar tracking angles are relative to the UCS.

RELATIVE TO LAST SEGMENT Polar tracking angles are relative to the last segment.

Section 8
Miscellaneous

BACKGROUND MASK

VERSION 2007, 2006 and 2005 only

Background mask inserts an opaque background so that objects under the text are covered. (masked) The mask will be a rectangular shape and the size will be controlled by the "Border offset factor". The color can be the same as the drawing background or you may select a different color.

1. Select the Multiline Text command.

2. Right click in the Text Area. (The dialog box below appear.)

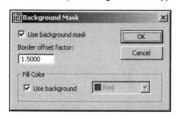

3. Turn this option **ON** by selecting the "Use background Mask" box.

4. Enter a value for the "Border offset factor". The value is a factor of the text height. 1.0 will be exactly the same size as the text. 1.5 (the default) extends the background by 0.5 times the text height. The width will be the same width that you defined for the entire paragraph.

5. In the Fill Color area, select the "Use background" box to make the background the same color as the drawing background. To specify a color, uncheck this box and select a color. (Note: if you use a color you may need to adjust the "draworder" using Tools / Draworder to bring the text to the front.)

BACK UP FILES

When you save a drawing file, Autocad creates a file with a **.dwg** extension.
For example, if you save a drawing as **12b**, Autocad saves it as **12b.dwg**. The next time you save that same drawing, Autocad replaces the old with the new and renames the old version **12b.bak**. The old version is now a back up file.
(Only 1 backup file is stored.)

How to view the list of back up files:
Select File / Open
Type "*.bak" in the "file name" box and <enter>. A list of the backup (.bak) files, within the chosen directory, will appear.

How to open a back up file:
You can't open a **.bak** file.
It must first be renamed with a **.dwg** file extension.

How to rename a back up file:
Right click on the file name.
Select "Rename". Change the .bak extension to .dwg and press <enter>.

GRIPS

Grips are little boxes that appear if you select an object when no command is in use. Grips must be enabled by typing "grips" <enter> then 1 <enter> on the command line or selecting the "enable grips" box in the **TOOLS / OPTIONS / SELECTION** dialog box.

Grips can be used to quickly edit objects. You can move, copy, stretch, mirror, rotate, and scale objects using grips.

The following is a brief overview on how to use three of the most frequently used options. Grips have many more options and if you like the example below, you should research them further in the AutoCAD help menu.

1. Select the object (no command can be in use while using grips)
2. Select one of the **blue** grips. It will turn to "**red**". This indicates that it is "**hot**". The "**Hot**" grip is the **basepoint**.
3. The editing modes will be displayed on the command line. You may cycle through these modes by pressing the SPACEBAR or ENTER key or use the shortcut menu.
4. <u>After editing you must press the ESC key to deactivate the grips on that object.</u>

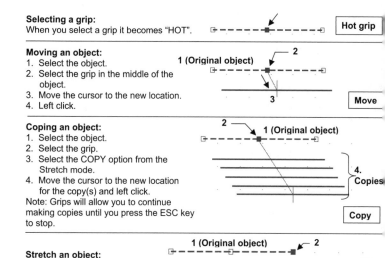

Selecting a grip:
When you select a grip it becomes "HOT".

| Hot grip |

Moving an object:
1. Select the object.
2. Select the grip in the middle of the object.
3. Move the cursor to the new location.
4. Left click.

1 (Original object)

| Move |

Coping an object:
1. Select the object.
2. Select the grip.
3. Select the COPY option from the Stretch mode.
4. Move the cursor to the new location for the copy(s) and left click.
Note: Grips will allow you to continue making copies until you press the ESC key to stop.

1 (Original object)

4. Copies

| Copy |

Stretch an object:
1. Select the object
2. Select the grip.
3. Move the cursor to stretch the object or type @X,Y

1 (Original object)

Stretch completed

| Stretch |

REMEMBER, press ESC key to deactivate the grips on an object.

OBJECT SNAP

Increment Snap enables the cursor to move in an incremental movement. So you could say your cursor is "snapping to increments" preset by you.

Object snap enables you to snap to "objects" in very specific and accurate locations on the objects. For example: The endpoint of a line or the center of a circle.

Selecting an Object Snap option using the Toolbar: Right click on any of the toolbars, then select the "Object Snap" toolbar from the list.

Selecting an Object Snap option using a Popup Menu:
Method 1: While holding down the shift key, press the right mouse button and the Object Snap menu will appear.

Method 2: Press the wheel and the Object Snap menu will appear.
Note: The command "Mbuttonpan" must be set to 0 for this option to function.)

OBJECT SNAP OPTIONS: *(Note: Refer to Lesson 5 for more Object Snap selections.)*

 ENDpoint — Snaps to the closest endpoint of a Line, Arc or polygon segment. Place the cursor on the object close to the end.

 MIDpoint — Snaps to the middle of a Line, Arc or Polygon segment. Place the cursor anywhere on the object.

 INTersection — Snaps to the intersections of any two objects. Place the Pick box directly on top of the intersection or select one object and then the other and Autocad will locate the intersection.

 CENter — Snaps to the center of an Arc, Circle or Donut. Place the cursor on the object, or the approximate center location.

QUAdrant — Snaps to a 12:00, 3:00, 6:00 or 9:00 o'clock location on a circle. Place the cursor on the circle near the desired quadrant location.

 PERpendicular — Snaps to a point perpendicular to the object selected. Place the cursor anywhere on the object.

	TANgent	Calculates the tangent point of an Arc or Circle. Place the cursor on the object as near as possible to the expected tangent point.
	NODe	This option snaps to a POINT object. Place the pick box on the POINT object.
	NEArest	Snaps to the nearest location on the nearest object. Select the object anywhere on the object.
	MTP	Locates a midpoint between two points
None		Snap to the first location and then the second. The midpoint between these 2 points will be located automatically. **Available in Versions 2007 & 2006 only.**

How to use OBJECT SNAP

The following is an example of attaching a line segment to previously drawn vertical lines. The new line will start from the upper endpoint (P1), to the midpoint (P2), to the lower endpoint (P3).

1. Select the Line command.
2. Draw two vertical lines as shown below.
3. Turn off "Increment Snap". (Use the Snap button or F9)
4. Select the Line command again.
5. Select the "Endpoint" object snap option using one of the methods listed on the previous page.
6. Place the cursor close to the upper endpoint of the left hand line (P1). *(Notice that a square appears at the end of the line. An "endpoint" tool tip should appear and the cursor snaps to the endpoint like a magnet. This is what "object snap" is all about. You are snapping the cursor to a previously drawn object.)*
7. Press the left mouse button to attach the new line to the endpoint of the previously drawn line. (Do not end the Line command yet.)
8. Now select the "Midpoint" object snap option.
9. Move the cursor to approximately the middle of the right hand vertical line (P2). A triangle and a "midpoint" tool tip appear, and the cursor should snap to the middle of the line like a magnet.
10. Press the left mouse button to attach the new line to the midpoint of the previously drawn line. (Do not end the Line command yet.)
11. Select the "endpoint" object snap option.
12. Move the cursor close to the lower endpoint of the left hand vertical line (P3).
13. Press the left mouse button to attach the new line to the endpoint of the previously drawn line.
14. Disconnect by pressing <enter>.

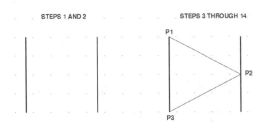

STEPS 1 AND 2   STEPS 3 THROUGH 14

RUNNING OBJECT SNAP

RUNNING OBJECT SNAP is a method of **presetting** the **object snap options**
so specific options, such as center, endpoint or midpoint, stay **active** until you
de-activate them. When Running Object Snap is active, markers are displayed
automatically as you move the cursor near the object and the cursor is drawn, to
the object snap location, like a **magnet**.

For example, if you need to snap to the endpoint of 10 lines, you could preset the
running object snap **endpoint** option. Then when you place the cursor near any
one of the lines, a marker will appear at the endpoint and the cursor will
automatically snap to the endpoint of the line. You then can move on to the next
and the next and the next. Thus eliminating the necessity of invoking the object
snap menu for each endpoint.

Running Object Snap can be toggled **ON** or **OFF** using the **F3** key or **clicking**
on the **OSNAP** button on the status bar.

<u>**Setting Running Object Snap**</u>
1. Select the **Running Object Snap** option using one of the following:

> **TYPE = OS <enter>**
> **PULL DOWN = TOOLS / DRAFTING SETTINGS**
> **Right Click on the OSNAP tile, on the Status Bar, and select**
> **SETTINGS.**

(The dialog box below will appear.)

2. Select the **OBJECT SNAP** tab.

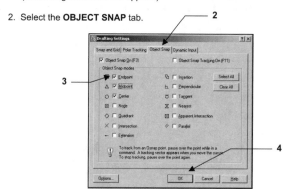

3. Select the Object Snap desired
4. Select **OK**.

Note: Do not preset more than 3 object snaps, you will lose control of the cursor.

PAN

After you adjust the scale of a viewport sometimes the drawing, within the viewport frame, is not placed as you would like it. If you use Zoom / All, the scale will not be correct and you would need to re-adjust the scale.

This is where the **PAN** command comes in handy. The PAN command will allow you to move the drawing around, within the viewport, without affecting the adjusted scale.

Note: Do not use the MOVE command. You do not want to actually move the original drawing. You only want to slide the viewport image, of the original drawing, around within the viewport.

How to use the PAN command.

1. Unlock the viewport.

2. Activate the viewport. (Double click inside viewport)

3. Select the **PAN** command using one of the following:

    TYPE = P
    PULLDOWN = VIEW / PAN / REALTIME
    TOOLBAR = STANDARD

4. Place the cursor inside the viewport and hold the left mouse button down while moving the cursor. (Click and drag) When the drawing is in the desired location release the mouse button.

5. Lock the viewport.
   Now you can use the Zoom commands and it will not affect the adjusted scale.

Note: You may set the "wheel" on your wheel mouse to Pan when you press and hold.
      Refer to "Customizing the Wheel mouse".

Example of before and after panning.

BEFORE PANNING

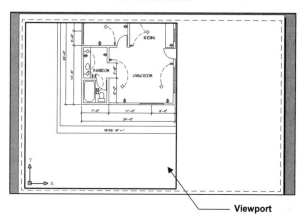

Viewport

AFTER PANNING

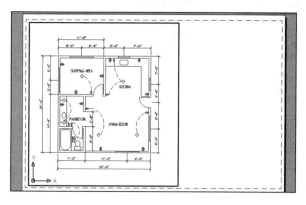

PROPERTIES PALETTE

The **Properties Palette**, shown below, makes it possible to change an object's properties. You simply open the Properties Palette, select an object and you can change any of the properties that are listed.

Open the Properties Palette by double clicking on an object, or use one of the following:

    TYPING = PROPS or CH
    PULLDOWN = MODIFY / PROPERTIES
    TOOLBAR = OBJECT PROPERTIES
    KEYS = CTRL + 1

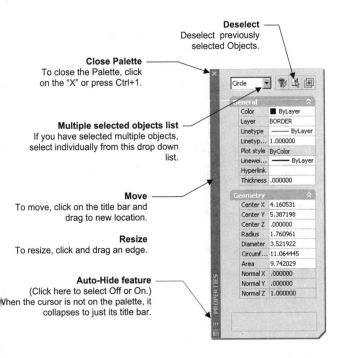

Deselect
Deselect previously selected Objects.

Close Palette
To close the Palette, click on the "X" or press Ctrl+1.

Multiple selected objects list
If you have selected multiple objects, select individually from this drop down list.

Move
To move, click on the title bar and drag to new location.

Resize
To resize, click and drag an edge.

Auto-Hide feature
(Click here to select Off or On.)
When the cursor is not on the palette, it collapses to just its title bar.

Note: The properties listed will depend on the object you have selected.

METRIC CONVERSION FACTORS

Multiply Length	By	To Obtain
centimeter	0.03280840	foot
centimeter	0.3937008	inch
foot	0.3048a	meter (m)
foot	30.48a	centimeter (cm)
foot	304.8a	millimeter (mm)
inch	0.0254a	meter (m)
inch	2.54a	centimeter (cm)
inch	25.4a	millimeter (mm)
kilometer	0.6213712	mile [U.S. statute]
meter	39.37008	inch
meter	0.5468066	fathom
meter	3.280840	foot
meter	0.1988388	rod
meter	1.093613	yard
meter	0.0006213712	mile [U. S. statute]
microinch	0.0254a	micrometer [micron] (mm)
micrometer [micron]	39.37008	microinch
mile [U.S. statute]	1609.344a	meter (m)
mile [U. S. statute]	1.609344a	kilometer (km)
millimeter	0.003280840	foot
millimeter	0.03937008	inch

Multiply Length	By	To Obtain
rod	5.0292a	meter (m)
yard	0.9144a	meter (m)

Area		
acre	4046.856	meter2 (m2)
acre	0.4046856	hectare
centimeter2	0.1550003	inch2
centimeter2	0.001076391	foot2
foot2	0.09290304a	meter2 (m2)
foot2	929.0304a	centimeter2 (cm2)
foot2	92,903.04a	millimeter2 (mm2)
hectare	2.471054	acre
inch2	645.16a	millimeter2 (mm2)
inch2	6.4516a	centimeter2 (cm2)
inch2	0.00064516a	meter2 (m2)
meter2	1550.003	inch2
meter2	10.763910	foot2
meter2	1.195990	yard2
meter2	0.0002471054	acre
mile2	2.5900	kilometer2
millimeter2	0.00001076391	foot2
millimeter2	0.001550003	inch2
yard2	0.8361274	meter2 (m2)
fathom	1.8288	meter (m)

Symbols of SI units, multiples and sub-multiples are given in parentheses in the right-hand column.

DRAWING SCALES

Scale	Drawing Scale Factor	Adjusted Scale times Paperspace
1/16=1'	192	1/192xp
3/32=1'	128	1/128xp
1/8=1'	96	1/96xp
3/16=1'	64	1/64xp
1/4=1'	48	1/48xp
3/8=1'	32	1/32xp
1/2=1'	24	1/24xp
3/4=1'	16	1/16xp
1=1'	12	1/12xp
1-1/2=1'	8	1/8xp
3=1'	4	1/4xp
1=10'	120	1/120xp
1=20'	240	1/240xp
1=25'	300	1/300xp
1=30'	360	1/360xp
1=40'	480	1/480xp
1=50'	600	1/600xp
1=60'	720	1/720xp
1=80'	960	1/960xp
1=100'	1200	1/1200xp
1=200'	2400	1/2400xp
1=10	10	1/10xp
1=20	20	1/20xp
1=16	16	1/16xp
1=30	30	1/30xp
1=40	40	1/40xp
1=50	50	1/50xp
1=100	100	1/100xp
2=1	0.50	2xp
4=1	0.25	4xp
8=1	0.125	8xp
10=1	0.10	10xp
100=1	0.01	100xp

PRINT A QUICK DRAFT ON A LETTER SIZE PRINTER.

Anytime you want to print a quick draft on your letter size printer do the following:

. Open the drawing on to the screen.

. Select **File / Plot**.

. Select your printer. (If your printer is not listed refer to Appendix A)

. The "AutoCAD Warning" box, shown below, will appear. Select the **OK** button.

. **Select** "Paper Size".
   Note: The paper size probably already changed to the default size for the printer.

. Select "**Extents**" for the Plot Area.

. Select "**Center** the Plot" for Plot Offset.

. Select "**Fit to Paper**" for Plot Scale.

. Select the "**Plot Style Table**".
   a. Black only = Monochrome
   b. Color = None (Of course your printer must be capable of printing in color.)

10. **Preview**
   a. If it looks correct press enter.
   b. If it doesn't look correct recheck 3 through 9 above.

Section 9
Plotting

BACKGROUND PLOTTING- Version 2007 & 2006 only

Background Plotting allows you to continue to work while your drawing is plotting. This is a valuable time saver because some drawings take a long time to plot. Or maybe you have multiple drawings to plot and you do not want to tie up your computer.

If you wish to view information about the status of the plot, click on the plotter icon located in the lower right corner of the AutoCAD window.

When the plot is complete, a notification bubble will appear.

 If you do not want this bubble to appear you may turn it off. Right click on the plotter icon and select "**Enable Balloon Notification**". This will remove the check mark and turn the notification off. You may turn it back on using the same process.

When you click on the "Click to view plot and publish details..." , on the bubble, you will get a report like the one shown below. The report will list details about all of the drawings plotted in the current drawing session.

 You may turn Background Plotting ON or OFF. The default setting is OFF.
To turn it ON or OFF:
select **Tools / Options / Plot and Publish tab.**

PLOTTING FROM MODEL TAB – Version 2004

1. **Important:** Open the drawing you want to plot.
2. Make sure that the Model tab is selected.   (Model tab)
3. Select the **Plot** command by "right clicking" on the "Model" tab or using one of the following methods listed below:

> **Type = Print or Plot**
> **Pulldown = File / Plot**
> **Tool bar = Standard**

The Plot dialog box below should appear.

4. Select the **"Plot Device"** tab.

5. Select the Printer (Plot device) such as the HP 4MV. **Note:** This printer represents a size 17 x 11. If you do not find your printer in the drop down list refer to "Configuring a Plotter".

   Notice: You may configure a printer even though your computer is not attached to it.

6. Select the Plot Style Table
   Select "**Monochrome.ctb**" to print black.

   Select "**None**" to print in color

7. Select the **"Plot Settings"** tab.

8. Check all the settings to make sure they match this example.

9. If you would like to print your drawing on a 8-1/2 X 11 printer, select the printer and change the scale to "Scaled to Fit".

10. Select **Full Preview** button.

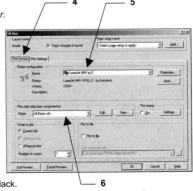

If you selected a different printer, these values may not be the same. That's OK.

9-3

11. If your drawing appears correct, press <enter>.

12. Select the **OK** button to send the drawing to the printer or <u>select **Cancel** if you do not want to print the drawing at this time or if you are not attached to this printer.</u>

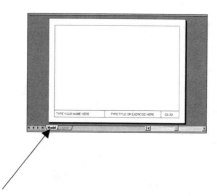

Note: The instructions above are to be used when plotting from the Model tab. To plot from a Layout tab refer to "Plotting from a Layout tab".

PLOTTING FROM MODEL TAB – 2007, 2006 & 2005

1. **Important:** Open the drawing you want to plot.
2. Make sure that the Model tab is selected.   (Model tab)
3. Select: View / Zoom / All
4. Select the **Plot** command by "right clicking" on the "Model" tab or using one of the following methods listed below:

> Type = Print or Plot
> Pulldown = File / Plot
> Tool bar = Standard

The Plot dialog box below should appear.

5. " More Options" button

5. Select the "**More Options**" button to expand the dialog box.

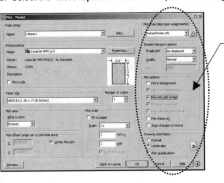

The dialog box is expanded to show more options

continue on to next page...

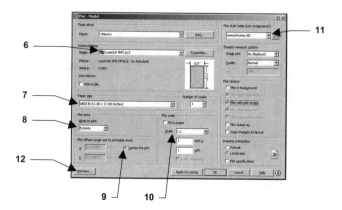

6. Select a "Printer / Plotter" such as the **LaserJet 4MV** from the drop down list.
   Note: This Printer represents a size 17 X 11. If you do not find your printer in the
   drop down list, you need to configure it on your system.
   Refer to "Configuring a Plotter".

7. Select the Paper Size such as **ANSI B (11 X 17 inches)**

8. Select the Plot Area **EXTENTS**

9. Select the Plot Offset **Center the plot**

10. Select Plot Scale **1 : 1**
    (Note: If you would like to print your drawing on a 8-1/2 X 11 printer, select the
    printer and change the scale to "Fit to Paper" and change paper size.)

11. Select the Plot Style table **None** for <u>color</u> or **Monochrome.ctb** for <u>Black only</u>.
    The following box will appear. Select **Yes**

Continue on to next page...

12. Select **Preview** button.

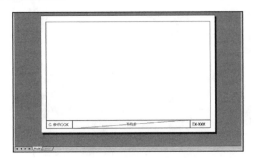

Does your display appear as correct?
If yes, press <enter> and proceed to 13.
If no, recheck 1 through 11.

You have just created a **Page Setup**. All of the settings you have selected can now be saved. You will be able to recall these settings for future plots using this page setup. To save the **Page Setup** you need to **ADD** it to the model tab within this drawing.

13. Select the ADD button.

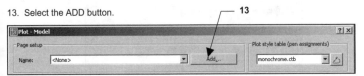

14. Type the new Page Setup name such as: **Model-mono**
(This name identifies that you will use it when plotting the Model tab in monochorme.)

15. Select **OK** button

continue on to next page....

16. Select **Apply to Layout** button.

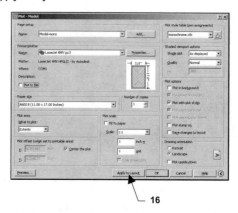

— 16

17. Select the **OK** button to send the drawing to the printer or <u>select **Cancel** if you do not want to print the drawing at this time. The Page Setup will still be saved.</u>

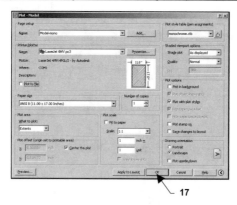

— 17

18. Save the entire drawing again. The Page Setup will be saved to the **Model tab** within the drawing and available to select in the future. You will not have to select all the individual settings.

Note: The instructions above are to be used when plotting from the "Model" tab. To plot from a Layout tab refer to Plotting from the Layout tab.

HOW TO PLOT A LAYOUT TAB - Version 2004 only

To plot a drawing from a Layout tab you should specify how you want it plotted.
Note: If you are selecting a Layout tab for the first time you must first create a page
setup. The you may return to this page. Refer to "How to create a page setup"

SPECIFY PLOT SETTINGS
Specifying Plot Settings means verify the Layout name, Plot device and paper size.
Select what to plot, scale and where to place the drawing on the paper.

A. Select **FILE / PLOT**

(The Plot dialog box shown below should appear)

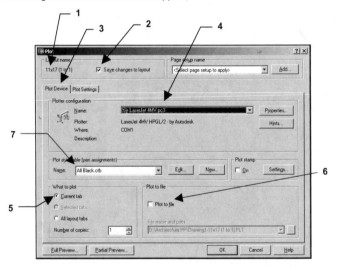

1. **Layout Name**: Displays the name of the Layout tab you selected.
   (To change the name of a layout, go back to paperspace, right click on the
   layout tab and select "Rename" from the short cut menu.)

2. **Save changes to Layout:** Check this box if you want your settings to be
   saved to the selected Layout tab.

3. Select the **"Plot Device"** tab.

4. Select the **"Plotter Name"** from the list. All previously configured devices are listed. (If your printer / plotter is not listed refer to "Add a Printer / Plotter" Appendix A.

5. **What to Plot:** defines what you want to plot.
   Current tab = plots current Model or Layout tab
   Selected tabs = plots multiple pre-selected tabs. This option is not available if only one tab is selected.
   All Layouts tabs = plots all layout tabs, selected or not.
   Number of copies = specify number of copies to be plotted.

6. **Plot to File:** Creates a plot file instead of plotting the drawing. If this option is selected, enter the **filename** and specify saving **location**. This is an advanced option, refer to the AutoCAD User's Guide for more information.

7. Select a **"Plot Style Table"** from the list.
   You can also **Create a New** plot style table

B. Select the **"Plot Settings"** tab.

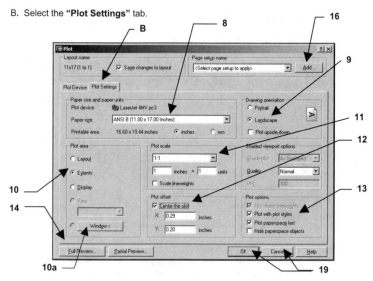

8. Select the desired **Paper Size** from the drop down list. The list should contain all available paper sizes for the plot device you selected.
   (Remember, you must select the plot device first)

9. Select the desired **Drawing Orientation**.
   Landscape = the long edge of the paper represents the top of the page.
   Portrait = the short edge of the paper represents the top of the page.
   (Landscape is the most frequently selected.)

9-10

10. Select the **Plot Area**.

Limits	plots the area inside the drawing limits.
	(Select when plotting from Modelspace)
Layout	plots the paper size (Select when plotting a Layout)
Extents	plots all objects in the drawing file even if out of view.
Display	plots the active viewport.
View	plots a previously saved view.
Window	plots objects inside a window. To specify the window, choose the

Window button **(10a)** and designate the first and opposite (diagonal) corner of the area you choose to plot. (Similar to the Zoom / Window command)

11. Select a **scale** from the drop down list or enter a custom scale.
*(If you are plotting from a "LAYOUT" tab, you should always use **plot scale 1:1**.*

12. **Plot Offset:** Specify where you want the drawing located on the sheet of paper. The **X** and **Y** boxes defines the offset from the lower left corner of the paper. The "**Center the Plot**" box automatically centers the drawing on the paper.

13. **Plot Options:**
Plot Object Lineweights = plots objects with assigned lineweights.
Plot with Plot Styles = plots using the selected Plot Style Table.
Plot paperspace last = plots model space objects before plotting paperspace objects. Not available when plotting from model space.
Hide Paperspace Objects = used for 3D only. Plots with hidden lines removed.

14. Select **Full Preview** button.
Full preview displays the drawing as it will plot on the sheet of paper. (Note: If you cannot see through to Model space, you have not cut your viewport yet)

a. If the drawing is centered on the sheet, press the **Esc** key and go on to step 15.

b. If the drawing does not look correct, press the **Esc** key and check all your settings, then preview again.

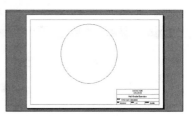

Notice the Viewport "frame" is not visible. The "Viewport" layer is set to "no plot"

NAME THE PAGE SETUP
After you have completed steps 1 through 4 you should save them by specifying a
name for the page setup.

16. Select the **"Add"** button

17. Type the new page set up name
    Try to give it a name that briefly describes the settings previously selected.

18. Select the **OK** button — **17**

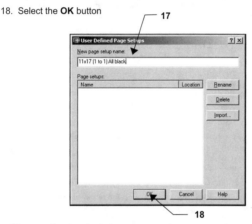

— **18**

The settings previously selected are now saved as a page setup

STEP 6. PLOT THE DRAWING or CLOSE THE PLOT DIALOG BOX.

19. *If your computer is connected to the plotter / printer*, select the **OK** button
    to plot, then proceed to step 7.
    If your computer is not connected to the plotter / printer, select the **Cancel**
    button to close the Plot dialog box and proceed to step 7. Note: Selecting Cancel
    <u>does not</u> cancel your page setup. It will still be saved to the drawing.

STEP 7. SAVE THE DRAWING AGAIN using **FILE / SAVE AS**. This will guarantee
    that the Page Setup you just created will be saved to this drawing for future
    use.

HOW TO PLOT FROM PAPER SPACE – 2007, 2006 & 2005

The previous page setup instructions were to select the printer and paper size.
Now you need to specify how you want to plot the drawing. You will find the PLOT
dialog box almost identical to the Page Setup dialog box.

STEP 1. SPECIFY PLOT SETTINGS

Specifying Plot Settings means:
Verify the Layout name, Plot device and Paper size.
Select what area of the drawing to plot, what scale to use, where to place the
drawing on the paper and which plot style table to use.

A. Select **FILE / PLOT.**

The Plot dialog box shown below should appear.
Select the "More Options" button (in the lower right corner) if your dialog box
does not appear the same as shown below.

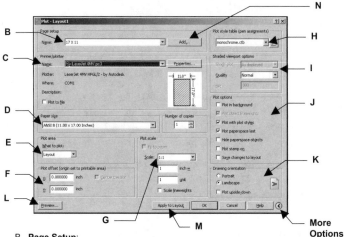

B. **Page Setup**:
Notice that the current page setup is displayed.
You may select a previously saved page setup from the drop down list.
If you make any changes to the settings, this name will change to <none>.

C. **Printer / Plotter:**
Verify the **Name.** All previously configured devices are listed.
If you would like to use your printer, selected it now instead of the
Hp4MV shown above.
(If your printer / plotter is not listed, refer to
"Add a Printer / Plotter")

D. **Paper Size:**
Verify the **Paper size**.

E. **Plot Area:**
Select the area to plot. Layout is the default.
Note: If you selected printer <u>HP4MV</u> select "Layout".
If you selected <u>your printer</u>, use "Extents".

Limits   plots the area inside the drawing limits.
          (Only shown when plotting from model space)
Layout   plots the paper size
          (Select when plotting a Layout)
Extents   plots all objects in the drawing file even if out of view.
          (Only shown if you have a viewport cut)
Display   plots the drawing as displayed.
Window   plots objects inside a window. To specify the window, choose
          Window and designate the first and opposite (diagonal) corner of
          the area you choose to plot. (Similar to the Zoom / Window
          command)

F. **Plot offset:**
The plot can be moved away from the lower left corner by changing the X
and/or Y offset.
If you select Plot area "Display" or "Extents", select "**Center the plot.**"

G. **Scale:** Select a **scale** from the drop down list or enter a custom scale.
Note: If you selected printer HP4MV, select scale 1 : 1.
If you selected your printer, select "<u>Fit to Paper size</u>".

*Note: This scale is the Paper Space scale. The Model space scale was
adjusted previously within the viewport. If you are plotting from a "LAYOUT"
tab, normally you will use plot scale 1:1.*

H. **Plot Style Table:** Select the Plot Style Table from the list.

I. **Shaded viewport options**
This area is used for printing shaded objects when using 3D.

J. **Plot options**
Plot Object Lineweights = plots objects with assigned lineweights.
Plot with Plot Styles = plots using the selected Plot Style Table.
Plot paperspace last = plots model space objects before plotting paperspace
objects. Not available when plotting from model space.
Hide Paperspace Objects = used for 3D only. Plots with hidden lines
removed.
Plot Stamp = Allows you to print information around the perimeter of the
border such as; drawing name, layout name, date/time, login name, device
name, paper size and plot scale.
Save Changes to Layout = Select this box if you want to save all of these
settings to the current Layout tab.

K. **Drawing Orientation**.
   Portrait = the short edge of the paper represents the top of the page.
   Landscape = the long edge of the paper represents the top of the page.

L. Select **Preview** button.
   Preview displays the drawing as it will plot on the sheet of paper.

(Note: If you cannot see through to Model space, you have not cut your viewport yet)

<u>If the drawing is centered</u> on the
sheet, press the **Esc** key and
continue.

<u>If the drawing does not look
correct,</u> press the **Esc** key and
check all your settings, then
preview again.

Note: It you set your Viewport layer to "No Plot" the Viewport "frame" will not appear
when you select Preview. What you see is what will print on the paper.

M. **Apply to Layout**
   This applies all of the previous settings to the layout tab. Whenever you
   select this layout tab the settings will be already set for you.

N. At this point you have the option of saving these settings as another page
   setup for future use, not just this layout tab. If you wish to save this setup,
   select the **ADD** button, type a name and select **OK.**

If your computer is connected to the plotter / printer selected, select the **OK**
button to plot, then proceed to **O**.

If your computer is __not__ connected to the plotter / printer selected, select the
Cancel button to close the Plot dialog box and proceed to **O**.

Note: Selecting Cancel <u>will cancel</u> your selected setting if you did not select the
"Apply to Layout" button or save the page setup as described in **N**.

O. Save the drawing
   This will guarantee that the Page Setup you just created will be saved to this
   file for future use.

Section 10
Settings

DRAFTING SETTINGS

The **DRAFTING SETTINGS** dialog box allows you to set the **INCREMENT SNAP** and **GRID SPACING**. You may change the Increment Snap and Grid Spacing at anytime while creating a drawing. The settings are only drawing aids to help you visualize the size of the drawing and control the movement of the cursor.

INCREMENT SNAP controls the movement of the cursor. If it is **OFF** the cursor will move smoothly. If it is **ON**, the cursor will jump in an *incremental* movement. This incremental movement is set by changing the **"Snap X and Y spacing"** .

GRID is the dot matrix in the drawing area. Grid dots will not print. The grid is only a visual aid. The Grid dot spacing is set by changing the **"Grid X and Y spacing".**

1. Select **DRAFTING SETTINGS** by using one of the following:

   TYPING = DS <enter>
   PULL-DOWN = TOOLS / DRAFTING SETTINGS
   TOOLBAR = NONE

2. The dialog box shown below will appear.

3. Select the **"Snap and Grid"** tab.

Note: The tabs *Object Snap* and *Dynamic Input* will be discussed in lessons 4 and 11.

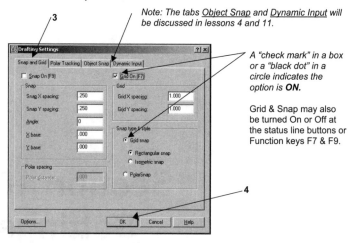

A "check mark" in a box or a "black dot" in a circle indicates the option is **ON.**

Grid & Snap may also be turned On or Off at the status line buttons or Function keys F7 & F9.

4. Make your changes and select the **OK** button to save them.
   If you select the **CANCEL** button, your changes will **not** be saved.

DRAWING LIMITS

DRAWING LIMITS

Consider the drawing limits as the size of the paper you will be drawing on. You will first be asked to define where the lower left corner should be placed, then the upper right corner, similar to drawing a Rectangle. An 11 x 17 piece of paper would have a **lower left corner** of 0,0 and an **upper right corner** of 17, 11. *(17 is the horizontal measurement or X-axis and 11 is the vertical measurement or Y-axis.)*

HOW TO SET THE DRAWING LIMITS

1. Select the **DRAWING LIMITS** command using one of the following:

   TYPE = LIMITS <enter>
   PULLDOWN = FORMAT / DRAWING LIMITS
   TOOLBARS = NONE

2. The following will appear on the command line:

   Command: '_limits
   Reset Model space limits:
   Specify lower left corner or [ON/OFF] <0.000,0.000>:

3. Type the X,Y coordinates **0, 0** for the lower left corner location of your piece of paper then press <enter>.

4. The command line will now read:

   Specify upper right corner <12.000,9.000>:

5. Type the X,Y coordinates **17, 11** for the upper right corner of your piece of paper then press <enter>.

6. **This next step is very important:** Select **VIEW / ZOOM / ALL** to make the screen display the new drawing limits.

LINETYPE SCALE

AutoCAD has many Linetypes. A Linetype is a series of dashes, lines and spaces. Each linetype has specific dimensions for the lines, dashes and spaces. When you are in paper space and you adjust the scale of the model space, AutoCAD automatically adjusts the scale of the linetype dashes, lines and spaces. So you do not have to do any modifications unless you desire.

Changing individual objects
If you want to change the linetype scale for an individual object use the Properties Palette.

Changing the entire drawing
If you want to change the linetype scale for the entire drawing:
1. On the command line type: *LTS <enter>*
2. LTSCALE Enter new linetype scale factor <1.0000>: *enter the new value*

NOTE:

If you are in Paperspace you will only need to change the value to 2 or .5 etc. AutoCAD attempts to adjust the linetype scale automatically to the drawing scale factor. But sometimes you want to tweek it a little.

Linetypes may not appear the same if you are in the model tab rather than the Layout tab. Linetype scale is handled differently in Model space.

Linetype scaling is a visual preference. There is no rule for spacing in cad.

PSLTSCALE
This variable controls paper space linetype scaling. This means that all linetype dash lengths and spaces will be scaled to the paper space scale.

1 = the viewports can have different adjusted scales and the linetype scale will be the same in all viewports. (this is the default setting)

0 = viewports with different adjusted scales will also appear to have differing linetype scales.

Note: Select View/Regen after changing Variable.

LINEWEIGHTS

It is "good drawing management" to draw related objects on the same layer. It is also "good drawing management" to establish a contrast in line weights between layers. For example, objects such as a house or a paper clip should be drawn with the "Object" layer and should have a greater line weight than the dimension layer or text layer.

The following are instructions for assigning Lineweights to Layers.

FIRST YOU NEED TO CHANGE THE LINEWEIGHT SETTINGS BOX.
1. Select **Format / Lineweight.**

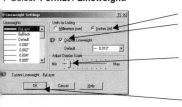

a. Select Inches or millimeters.
b. Select "Display Lineweight" box
   (you may use status line button [LWT])

c. Slide "Adjust Display Scale" to the left as shown. (Controls Lineweight appearance on the screen only)

d. Select OK

These settings will be saved to the computer not the drawing and will remain until you change them.)

ASSIGNING LINEWEIGHTS TO LAYERS
2. Select **Format / Layer**

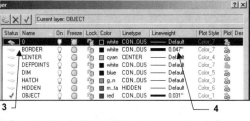

Notice the 0.047" Lineweight setting for the "Border" layer and the 0.031" Lineweight setting for the "Object" layer.

Default = 0.010

3. Select the "Border" layer. (Click on the name "Border")

4. Click on the Lineweight.

5. Scroll and select from the list, then OK.

*(Lineweight changes will be saved **only to the current drawing**, if you save it, and will not affect any other drawing)*

SETTING THE PICK BOX SIZE

☐ Pick box

When AutoCAD prompts you to **select objects,** such as when you are erasing objects, the cursor (crosshairs) turns into a square. This square is called a **Pick box.** The size of the **Pick box** can be changed. Some AutoCAD users prefer large boxes, some like small boxes. The size of the box is your personal preference, however, the smaller the Pick box the more **accurate** you must be when placing the pick box on an object to select it. If the Pick box is too large it could overlap onto other objects that you did not want to select.

HOW TO CHANGE THE "PICK BOX" SIZE.

1. Select **TOOLS / OPTIONS**
2. Select the **SELECTION** tab.

The following dialog box will appear.

3. Adjust the size by sliding the tab (click and drag) to the right (max) or to the Left (Min.). A preview of the new size is displayed.

2

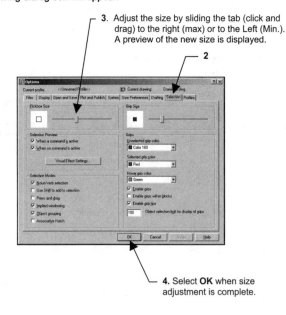

4. Select **OK** when size adjustment is complete.

UNITS and PRECISION

UNITS AND PRECISION
It is necessary to select what unit of measurement you want to work with.
Such as: Decimal (0.000) or Architectural (0'-0").
Next you should select how precise you want the measurements. This means, do you
want the measurement rounded off to a 3 place decimal or the nearest 1/8".

HOW TO SET THE UNITS AND PRECISION.
1. Select the **UNITS** command using one of the following:

> TYPE = UNITS <enter>
> PULLDOWN = FORMAT / UNITS
> TOOLBAR = NONE

(The dialog box below will appear.)

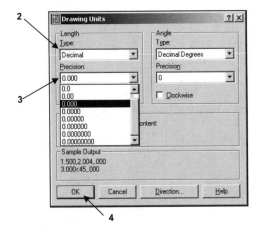

2. Select the appropriate **TYPE** such as: decimals or architectural.

3. Select the appropriate **PRECISION** associated with the "type".

4. Select the **OK** button to save your selections.

Section 11
Text

CREATING NEW TEXT STYLES

AutoCAD provides you with only one Text Style named "Standard". You may want to create a new text style with a different font and effects. Steps 1 through 8 below will guide you through the process.

1. Select the TEXT STYLE command using one of the following:
   TYPE = STYLE or ST
   PULLDOWN = FORMAT / TEXT STYLE
   TOOLBAR = FORMAT

   The TEXT STYLE dialog box below should appear.

The information in this dialog box is a description of the Text Style highlighted in the Style Name box.

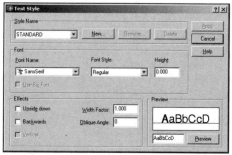

2. Select the **NEW** button.

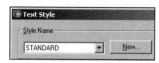

3. Type the new style name in **STYLE NAME** box. Then select the **OK** button.

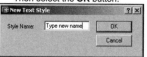

Text Styles can have a maximum of 31 characters, including letters, numbers, dashes, underlines and dollar signs. You can use Upper or Lower case.

4. Select the **FONT**.

5. Enter the value of the Height.
   (**If the value is 0, AutoCAD will always prompt you for a height. If you enter a number the new text style will have a fixed height and AutoCAD will not prompt you for the height**)

Assign **EFFECTS.**

‍PSIDE-DOWN
‍ach letter will be created upside-down in the order in which it was typed.
‍Note: this is different from rotating text 180 degrees.)

‍ACKWARDS
‍he letters will be created backwards as typed.

‍ERTICAL
‍ach letter will be inserted directly under the other. Only **.shx** fonts can be used.
‍ERTICAL text will not display in the **PREVIEW** box.

‍BLIQUE ANGLE
‍reates letter with a slant, like italic. An angle of 0 creates a vertical letter. A positive angle
‍ill slant the letter forward. A negative angle will slant the letter backward.

‍IDTH FACTOR
‍his effect compresses or extends the width of each character. A value less than 1
‍ompresses. A value greater than 1 extends each character.

. **PREVIEW** your settings.

. Select the **APPLY** button and then the **CLOSE** button.

CREATING YOUR NEW TEXT STYLE IS NOW COMPLETE.

CHANGING TEXT STYLES

RENAMING

1. Select the **TEXT STYLE** command.
   The Text Style Dialog box appears.

2. Select the style you want to rename.

3. Click on **RENAME** button.

4. Type the New name then click on the **OK** button.

5. Click on the **CLOSE** button.

DELETING

1. Select the **TEXT STYLE** command.
   The Text Style Dialog box appears.

2. Select the style you want to DELETE.

3. Click on the **DELETE** button.

4. Warning appears, select **Yes or No**.

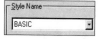

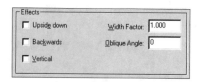

5. Click on the **CLOSE** button.

CHANGING EFFECTS

1. Select the **TEXT STYLE** command.
   The Text Style Dialog box appears.

2. Make the changes.

Effects	
☐ Upside down	Width Factor: 1.000
☐ Backwards	Oblique Angle: 0
☐ Vertical	

3. Click on the **APPLY** button and then **CLOSE**.

MULTILINE TEXT or MText

MULTILINE TEXT command allows you to easily add a sentence, paragraph or tables. The Mtext editor has most of the text editing features of a word processing program. You can underline, bold, italic, add tabs for indenting, change the font, line spacing, and width of the paragraph.

When using MText you must first define a text boundary box. The text boundary box is defined by entering where you wish to start the text (first corner) and approximately where you want to end the text (opposite corner). It is very similar to drawing a rectangle. The paragraph text is considered one object rather than several individual sentences.

USING MULTILINE TEXT

1. Select the MULTILINE TEXT command using one of the following:

    TYPE = MT
    PULLDOWN = DRAW / TEXT / MULTILINE TEXT
    TOOLBAR = DRAW [A]

    The command line will lists the current style and text height. The cursor will then appear as crosshairs with the letters "abc" attached. These letters indicate how the text will appear using the current font and text height.

    Mtext Current text style: "STANDARD"   Text height: .250

2. Specify first corner: *Place the cursor at the upper left corner to start the new text boundary box and press the left mouse button. (P1)*

3. Specify opposite corner or [Height / Justify / Line Spacing / Rotation / Style / Width]: *Move the cursor to the right and down (P2) and press left mouse button.*

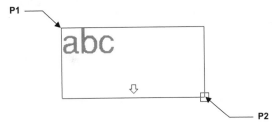

The (2 piece) In-Place Text Editor will appear.

Version 2007 & 2006

Text Formatting Tool Bar

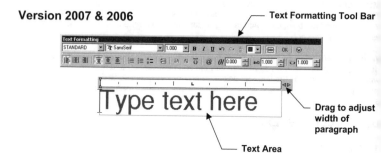

Drag to adjust width of paragraph

Text Area

Version 2005 and 2004

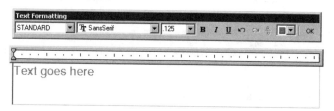

Notice the **In-Place Text Editor** is in 2 pieces: The **Text Formatting toolbar** and the **Text Area**. The toolbar portion can be moved but the text area remains in the location that you designated.

The **Text Formatting toolbar** allows you to select the Text Style, Font, Height etc. You can add features such as bold, italics, underline and color. There is even an UNDO button.

The **Text Area box** allows you to enter the text, add tabs, adjust left hand margins and change the width of the paragraph.

4. After you have entered the text in the Text Area box, do one of the following to add the new text to the drawing and close the Mtext Editor:
   a. Select the OK button
   b. Press Ctrl + <enter>
   c. Left Click anywhere outside the Mtext Editor, but within the drawing area.

MTJIGSTRING

HOW TO CHANGE THE "abc", ON THE CROSSHAIRS, TO OTHER LETTERS.

You can personalize the letters that appear attached to the crosshairs using the **MTJIGSTRING** system variable. (10 characters max) The letters will simulate the appearance of the font and height selected but will disappear after you place the lower right corner (P2).

1. Type **MTJIGSTRING** <enter> on the command line.
2. Type the new letters <enter>.

The letters will be saved to the computer, not the drawing. They will appear anytime you use Mtext and will remain until you change them again.

LINE SPACING for Multiline Text

The Multiline text command allows you to set the spacing between the bottom of the first line of text to the bottom of the following lines of text. This is accomplished using the **Line spacing** option within the MText command.

You may set the Line spacing to a **Factor** of the "original line spacing", or enter the **Specific** distance desired.

If you choose to use the Factor method, you may use a factor up to 4X the "original line spacing". AutoCAD has established the "original line spacing" as 1.66 of the text height. For example, the line spacing for 1" text is 1.66 from the bottom of the first line to the bottom of the second line.

Line spacing_____

(Original line spacing distance is 1.66 X Text ht.)

Line spacing_____ ▼

If you choose to enter a specific distance, the original line spacing is ignored.

Set the Line spacing.

1. Select the MText command.

   Command: _mtext Current text style: "STANDARD" Text height: 1.00
2. Specify first corner: *Place the cursor, and left click, to locate the first corner of the text boundary.*

3. Specify opposite corner or [Height/Justify/Line spacing/Rotation/Style/Width]: *Select the Line spacing option by typing L <enter> or right click and select Line spacing from the short cut menu.*

4. Enter line spacing type [At least/Exactly] <At least>: *Select " Exactly" option by typing E <enter> or right click and select "Exactly" from the short cut menu.* ("At least" is the default option that merely insures that the text does not overlap)

5. Enter line spacing factor or distance <1x>: *Enter a factor or a specific distance. (Note: you must include the "x" when entering the factor number.)*

6. Specify opposite corner or [Height/Justify/Line spacing/Rotation/Style/Width]: *Place the cursor, and click, to locate the opposite corner of the text paragraph.*

   The MText editor will appear.

7. Type the text and select OK.

Edit the Line spacing
To edit the existing line spacing use the Properties Palette.

EDITING MULTILINE TEXT

Multiline Text is as easy to edit as it is to input originally. You may change the style, font, height, color, indent and add text features such as bold, italic and underline.

1. Double click on the Multiline text you want to edit.

The "In-Place Editor" will appear:

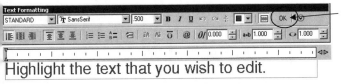

2. Highlight the text, that you want to change, using click and drag.

3. Make the changes then select the **OK** button.

> **Note: The "In-Place Editor" above is for Version 2006. Refer to Multiline Text to view Version 2005 and 2004 "Text Editor".**

You may edit many other Multiline Text features by right clicking in the Text Area and selecting an option from the menu shown below.

INDENTS

Sliders on the ruler show indention relative to the left side of the text boundary box. The top slider indents the first line of the paragraph, and the bottom slider indents the other lines of the paragraph.

You may change their positions at anytime, using one of the following methods.

Method 1.
Place the cursor on the "Slider" and click and drag it to the new location.

Method 2.
1. Right click on the Ruler.
2. Select "Indents and Tabs" from the short cut menu.

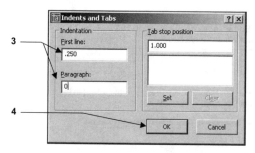

3. Type the indent position in the First line and/or Paragraph box.
4. Select OK button.

First Line:
Sets indentation for the first line of the current paragraph or selected paragraphs. (The top slider)

Paragraph:
Sets indentation for the current paragraph or selected paragraphs. (The lower slider)

TABS

Setting and removing Tabs is very easy.

The default setting for tabs is 1". (You may set as many tabs as you need.)
Set or change the stop positions at anytime, using one of the following methods.

Method 1.
Place the cursor on the "Ruler" where you want the tab and left click. A little dark "**L**" will appear. The tab is set.
If you would like to remove a tab, just click and drag it off the ruler and it will disappear.

TAB

Method 2.
1. Right click on the Ruler.
2. Select "Indents and Tabs" from the short cut menu.

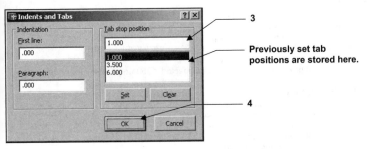

3

Previously set tab
positions are stored here.

4

3. Set a tab position by typing the position in the upper box, then select the "Set" button.
4. Select the OK button.

Tab Stop Position
Sets tab positions for the current paragraph or selected paragraphs.
The list below the text box shows the current tab stops.

Set
Copies the value in the Tab Stop Position box to the list below the box.

Clear
Removes the selected tab stop from the list.
Clear a tab position by highlighting the position in the lower box then select the "Clear" button.

SCALING TEXT

The **Scaletext** command allows you to scale <u>Single or Multiline</u> text using **height**, **factor** or **match** a previously drawn text. The text will be scaled proportionately in the X and Y axis. In other words, it gets larger or smaller all over. You must select a "base point" from which the text will enlarge or reduce. The base point is a justification option. If you would like it to scale from the middle, you will select "middle" as the base point. If you would like the text to scale from the top, you will select "TC". Most of the time you will simple use "existing". Existing is the justification used when the text was originally created.

Select the **SCALETEXT** command using one of the following:

> **TYPE = scaletext**
> **PULLDOWN = MODIFY / OBJECT / TEXT / SCALE**
> **TOOLBAR = TEXT**

HEIGHT
1. Select the line of text you want to edit.
2. Select objects: *select more text or <enter> to stop selecting*
3. Enter a base point option for scaling
   [Existing/Left/Center/Middle/Right/TL/TC/TR/ML/MC/MR/BL/BC/BR]
   <Existing>:*select the base point*
4. Specify new height or [Match object/Scale factor] <.250>:*type new height <enter>*

MATCH OBJECT
1. Select the line of text you want to edit.
2. Select objects: *select more text or <enter> to stop selecting*
3. Enter a base point option for scaling
   [Existing/Left/Center/Middle/Right/TL/TC/TR/ML/MC/MR/BL/BC/BR]
   <Existing>:*select the base point*
4. Specify new height or [Match object/Scale factor] <.250>:*select Match object*
5. Select a text object with the desired height: *select the text to match*
   Height = *the new height will be shown here*

SCALE FACTOR
1. Select the line of text you want to edit.
2. Select objects: *select more text or <enter> to stop selecting*
3. Enter a base point option for scaling
   [Existing/Left/Center/Middle/Right/TL/TC/TR/ML/MC/MR/BL/BC/BR]
   <Existing>:*select the base point*
4. Specify new height or [Match object/Scale factor] <.250>:*select Scale factor*
5. Specify scale factor or [Reference] <2.000>: *type the factor*

SINGLE LINE TEXT

SINGLE LINE TEXT allows you to draw one or more lines of text. The text is visible as you type. To place the text in the drawing, you may use the default **START POINT** (the lower left corner of the text), or use one of the many styles of justification described on the next page.

USING THE DEFAULT START POINT

1. Select the **SINGLE LINE TEXT** command using one of the following:

TYPE = DT or TEXT
PULLDOWN = DRAW / TEXT / SINGLE LINE TEXT
TOOLBAR = DRAW A̲

    Command: _dtext
    Current text style: "STANDARD" Text height: 0.250
2. Specify start point of text or [Justify/Style]: *Place the cursor where the text should start and left click.*
3. Specify height <0.250>: *type the height of your text*
4. Specify rotation angle of text <0>: *type the rotation angle then <enter>*
5. Enter text: *type the text string; press enter at the end of the sentence*
6. Enter text: *type the text string; press enter at the end of the sentence*
7. Enter text: *type the next sentence or press <enter> to stop*

USING JUSTIFICATION

If you need to be very specific, where your text is located, you must use the Justification option. For example if you want your text in the middle of a rectangular box, you would use the justification option "Middle".

The following is an example of Middle justification.

1. Draw a Rectangle 6" wide and 3" high.
2. Draw a Diagonal line from one corner to the diagonal corner.
3. Select the SINGLE LINE TEXT command
    Command: _dtext
    Current text style: "STANDARD" Text height: 0.250
4. Specify start point of text or[Justify/Style]: *type "J"*
5. Enter an option [Align/Fit/Center/Middle/Right/TL/ TC/TR/ML/MC/MR/BL/BC/BR]: *type M*
6. Specify middle point of text: *snap to the midpoint of the diagonal line*

7. Specify height <0.250>: *1 <enter>*
8. Specify rotation angle of text <0>: *0 <enter>*
9. Enter text: *type: HHHH <enter>*
10. Enter text: *press <enter> to stop*

OTHER JUSTIFICATION OPTIONS:

ALIGN
Aligns the line of text between two points specified.
The height is adjusted automatically.

FIT
Fits the text between two points specified.
The height is specified by you and does not change.

CENTER HyyHHyyHHyy
This is a tricky one. Center is located at the bottom center of Upper Case letters.

MIDDLE ┼HHHH╫HHHH┼   HHyy╫HyyHHyy
If only uppercase letters are used: located in the middle, horizontally and vertically.
If both uppercase and lowercase letters are used: located in the middle, horizontally and
vertically, of the lowercase letters.

RIGHT HyyHHyyHHyy
Bottom right of upper case text.

TL, TC, TR HyyHHyyHHyy
Top left, Top center and Top right of upper and lower case text

ML, MC, MR HyyHHyyHHyy
Middle left, Middle center and Middle right of upper case text.
(Notice the difference between "Middle" and "MC"

BL, BC, BR HyyHHyyHHyy
Bottom left, Bottom center and Bottom right of lower case text.

EDITING SINGLE LINE TEXT

SINGLE LINE TEXT

Editing **Single Line Text** is somewhat limited compared to Multiline Text. In the example below you will learn how to edit the text within a Single Line Text sentence.

VERSION 2007 & 2006
1. Double click on the Single Line text you want to edit. The text will highlight.

2. Make the changes in-place then press <enter><enter>.

VERSION 2005 and 2004
1. Double click on the Single Line text you want to edit.

The following "Edit Text" box will appear:

2. Make the changes then select the OK button.

3. Select the next line of text to be edited or press <enter> to stop

SPECIAL TEXT CHARACTERS

Characters such as the ***degree symbol, diameter symbol*** and the ***plus / minus symbols*** are created by typing **%%** and then the appropriate "code" letter.
For example: entering 350**%%D** will create: **350°**. The **"D"** is the **"code"** letter.

SYMBOL	CODE
∅ Diameter	%%C
° Degree	%%D
± Plus / Minus	%%P

SINGLE LINE TEXT
If you are using "Single Line Text", type the code in the sentence. While you are typing, the code will appear, in the sentence, on the command line. But when you are finished typing, and press <enter>, the symbol will appear.

MULTILINE TEXT
The code displays the same as single line text if you type your text in the text editor. But multiline text offers another method for inserting special character "symbols".

1. While you are typing in the in-place text editor, instead of typing the code, right click and the menu will appear.

2. Select "Symbol".

3. Select the symbol desired and the code or the actual symbol will automatically appear in the sentence.

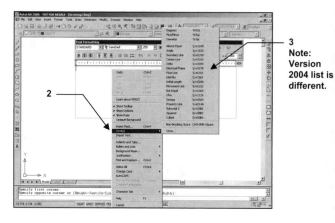

3
Note:
Version
2004 list is
different.

Section 12
UCS and Origin

DISPLAYING THE UCS ICON

The UCS icon is merely a drawing aid. You control how it is displayed.
It can be visible (on) or invisible (off). It can move with the Origin or stay in the default location. You can even change its appearance.

Select the following pull-down menu:

View / Display / UCS Icon

ON: A check mark beside the word ON means the UCS icon will be visible.
Remove the check mark and the UCS icon will disappear.
It will be very helpful to have the UCS Icon "On" most of the time. It displays where the Origin is located.

ORIGIN: A check mark beside the word Origin will move the UCS Icon with the Origin each time you move the Origin. Remove the check mark and the icon will not move.

PROPERTIES: This setting allows you to change the appearance of the UCS icon.
When you select this option the dialog box shown below will appear.
You may change the Style, Size and Color at any time. Changing the appearance is personal preference and will not affect the drawing or commands.

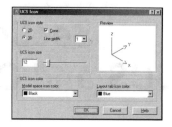

MOVING THE ORIGIN

The **ORIGIN** is where the X, Y, and Z axes intersect. The Origin's (0,0,0) default location is in the lower left-hand corner of the drawing. But you can move the Origin anywhere on the screen using the UCS command.
(The default location is designated as the "**World**" option or WCS. When it is moved it is UCS.)

You may move the Origin many times while creating a drawing. This is not difficult and will make it much easier to draw objects in specific locations.

To MOVE the Origin:

1. Select one of following:

        TYPING = UCS <enter> M <enter>   **(2007) = None**
        PULLDOWNS = TOOLS / Move UCS   **(2007) = TOOLS / New UCS / Origin**
        UCS II TOOLBAR =

Command: _ucs
Current ucs name: *World*
Enter an option [New/Move/orthoGraphic/Prev/Restore/Save/Del/Apply/?/World]
<World>: _move

2. Specify new origin point or [Zdepth]<0,0,0>: *type coordinate or use the cursor to place.*

To RETURN the Origin to the default "World" location (the lower left corner):

1. Select one of the following:

        TYPING = UCS <enter> W <enter>
        PULLDOWNS = TOOLS / New UCS / World
        UCS TOOLBAR =

INDEX

A

Absolute Coordinate	7-3
Add a Printer	5-2
Aligned, Dimension	3-6
Angle, Chamfer	1-9
Angular, Dimension	3-7
Arc	4-2
Arc length, Dimension	3-8
Array	1-2
Array, Polar	1-3
Array, Rectangular	1-2
Arrow with Line	3-16
Arrowheads, Dim.	3-45
Assign Lineweights to color	5-27
Associative Dim's	3-2

B

Background Mask	8-2
Background Plotting	9-2
Backup Files	8-3
Baseline, Dimension	3-5
Block Layers	4-9
Block, Creating a	4-6
Block, Dynamic	4-14
Block, Editing a	4-13
Block, Inserting	4-10
Block, Purging a	4-12
Block, Re-defining a	4-12
Bmake	4-6
Break	1-6

C

Calculate DSF	5-23
Center, Object Snap	8-5
Center Mark	4-25
Center Marks, Dim.	3-45
Chamfer	1-8
Chamfer, Angle	1-9
Chamfer, Distance	1-8

Characters, Special Text	11-16
Circle	4-26
Configure a Printer	5-2
Continue, Dimension	3-5
Convert Object to Rev. Cloud	4-48
Color, Layer	6-7
Compare 2 Dim. Styles	3-40
Conversion Factors, Metric	8-12
Coordinate, Absolute	7-3
Coordinate, Polar	7-11
Coordinate, Relative	7-3
Coordinate Input	7-2
Coordinate Input examples	7-4
Copy	1-10
Create new Layer (2004)	6-5
Create new Layer (06 & 05)	6-6
Create a Page Setup	
2004	5-6
2007, 2006 and 2005	5-9
Create a Table	5-33
Create a Template	5-13
Creating a Text Style	11-2
Curve, Dimension large	3-10
Customize wheel mouse	5-22

D

Datum feature symbol	3-25
Datum Triangle	3-26
DDE	7-5
Deviation, Tol. Dim.	3-22
Diameter dimension	3-11
Display UCS icon	12-2
Dim, Aligned	3-6
Dim, Alternate Units	3-43
Dim, Angular	3-7
Dim, Arc length	3-8
Dim, Associative	3-2
Dim, Baseline	3-4
Dim, Continue	3-5
Dim, Diameter	3-11

Dim, Edit Properties	3-29
Dim, Edit Text Value	3-30
Dim, Edit Text Position	3-32
Dim, Flip	3-12
Dim, Geometric Tolerance	3-23
Dim, Large Curve	3-10
Dim, Leader	3-15
Dim, Leader, Geo. Tol.	3-24
Dim, Leader, Hookline	3-15
Dim, Linear	3-4
Dim, Metric	3-43
Dim, Ordinate	3-17
Dim, Override	3-42
Dim, Qdim	3-13
Dim, Qdim editing	3-33
Dim, Quick	3-13
Dim, Qleader	3-15
Dim, Radial	3-11
Dim, Radius	3-12
Dim, Reassociate	3-3
Dim, Tolerance	3-21
Dimension Feature Scale	3-47
Dimension Lines	3-44
Dim. Measurement scale	3-48
Dim. Style, Create New	3-36
Dim. Style, **Create Sub-Style**	3-41
Dim. Style Definitions	3-44
Dim. Style Manager	3-35
Dim. Style, Modify	3-28
Dim. Trans-spatial	3-49
Direct Distance Entry	7-5
Distance, Chamfer	1-8
Distance, Inquiry	1-17
Divide	1-11
Donut	4-28
Drafting Settings	10-2
Draw something big	2-3
Draw something small	2-5
Drawing Limits	10-3
Drawing Scale Factor..	
How to calculate	5-23
How effects dimensions	5-24
How effects hatch	5-25
How effects text	5-26
Drawing scales	8-14
Dynamic Block	4-15
2007	4-20
2006	4-16
Dynamic Input	7-6

E

Extension Lines, Dim.	3-44
Edit a block	4-13
Edit Dimension Properties	3-29
Edit Dim. Text Value	3-30
Edit Dim. Text Position	3-32
Edit Hatch	4-36
Edit Polyline	4-43
Edit Qdim	3-33
Edit Single Line Text	11-15
Edit Text Style	11-4
Ellipse	4-29
Endpoint, Object Snap	8-5
Erase	1-12
Exit AutoCAD	5-17
Explode	1-13
Extend	1-14

F

Field, Create a	5-41
Field, Add to table cell	5-43
Field, Edit a	5-43
Field, Update a	5-42
Files, Back up	8-3
Fill mode	4-41
Fillet	1-15
Fine Tuning, Dim.	3-47
Fit Options, Dim.	3-47
Flip	3-12
Freeze and Thaw (2004)	6-10
Freeze and Thaw (06 & 05)	6-11

G

GDT.shx Text	3-27
Geometric Symbol Text	3-27
Geometric Tolerance	3-23
Gradient Fills	4-30
Grips	8-4

H

Hatch – 2005/2004	4-32
Hatch, Associative	4-34
Hatch, Gap Tolerance	4-35
Hatch, Predefined	4-33
Hatch, Preview	4-35
Hatch, User defined	4-34
Hatch – 2007/2006	4-36
Hatch, Associative	4-38
Hatch, Gap Tolerance	4-39
Hatch, Predefined	4-37
Hatch, Preview	4-40
Hatch, User defined	4-38
Hatch, Editing	4-40
Hatch, Ignor	4-41
Hatch, Trimming	4-40
Hookline, Leader	3-15
How to…	
Calculate DSF	5-23
Customize mouse	5-22
Insert a Table	5-35
Lock a Viewport	5-21
Save a drawing	5-30
How to Create..	
Page Setup – 2004	
Table	5-33
Viewport	5-18
How to use..	
Object Snap	8-7
Running Object Snap	8-8

I

Icon, Display UCS	12-2
ID Point, Inquiry	1-17
Ignor Hatch	4-37
Input, Coordinate	7-3
Input, Dynamic	7-6
Input, Polar Coordinate	7-11
Inquiry	1-17
Inquiry, Distance	1-17
Inquiry, ID Point	1-17
Inquiry, List	1-17
In-Place Text editor	11-6
Indents	11-10
Intersection, Object Snap	8-5

J

Jog, Ordinate dimension	3-19
Justify Text	11-14

K

L

Layer, Color	6-7
Layer, Selecting a	6-2
Layer, Selecting a (2004)	6-4
Layer, Selecting a (06 & 05)	6-3
Layer Control Definitions (04)	6-10
Layer Control Definitions (06)	6-11
Layer, Freeze a (2004)	6-10
Layer, Freeze a (06 & 05)	6-11
Layer, Plot/no Plot	6-11
Layers	6-2
Layout tab	2-2
Layout tab, plotting from (04)	9-9

Layout tab, plotting from (06) 9-13
Leader 3-15
Leader, Geo. Tol. 3-24
Limits, Drawing 10-3
Limits, Tol. Dim. 3-22
Line, Draw a 4-38
Line with Arrow only 3-16
Linear, dimensions 3-4
Linespacing for Mtext 11-8
Linetype, Loading a (2004) 6-8
Linetype, Loading a (06 & 05) 6-8
Linetype Scale 10-4
Lineweights 10-5
Lineweights to color, Assign 5-27
List, Inquiry 1-17
Loading a Linetype (2004) 6-8
Loading a Linetype (06 & 05) 6-9
Lock a Layer (2004) 6-10
Lock a Layer (06 & 05) 6-11
Lock a Viewport 5-21

M

Match Properties 1-18
Mbuttonpan 5-22
Measure 1-19
Measurement scale, Dim. 3-48
Metric Conversion Factors 8-12
Midpoint, Object Snap 8-5
Mirror 1-20
Mirrtext 1-20
Model tab 2-2
Model tab, plotting from (04) 9-3
Model tab, plotting from (06) 9-5
Model space 2-2
Modify, Dim. Style 3-28
Mouse, Customize a 5-22
Move 1-21

Moving the Origin 12-3
MTP, Object Snap 8-6
MTJIGSTRING 11-7
Multiline Text 11-5
Multiline Text Editing 11-9
Multiline Tex Indents 11-10
Multiline Text Line spacing 11-8
Multiline Text Tabs 11-11

N

Nearest, Object Snap 8-6
New drawing, Start a 5-32
Node, Object Snap 8-6

O

Object Snap 8-5
Object Snap, How to use 8-7
Object Snap, Running 8-8
Offset 1-22
Offset options 1-23
Open an Existing dwg 5-16
Open a Template 5-15
Ordinate dimension 3-17
Ordinate dimension jog 3-19
Origin, Moving the 12-3

P

Page setup 5-6
Palette, Properties 8-11
Pan 8-9
Paper space 2-2
Perpendicular, Object Snap 8-5
Pick Box size 10-6
Plotting
    Background (2006) 9-2
    From Layout tab (2004) 9-9
    From Layout tab (06/05) 9-13

From model tab (2004)	9-3
From model tab (06/05)	9-5
Point	4-39
Point Style	4-39
Polar Array	1-3
Polar Coordinate Input	7-11
Polar Snap	7-12
Polar Tracking	7-13
Polygon	4-40
Polyline	4-41
Polyline, Editing a	4-43
Polyline Options	4-42
Precision, Setting the	10-7
Printer, Add a	5-2
Properties Palette	8-11
Properties, Edit Dimension	3-29
Properties, Match	1-18
Purge a block	4-12

Q

Qdim	3-13
Qdim, Baseline	3-13
Qdim, Continuous	3-13
Qdim, Diameter	3-14
Qdim, Radius	3-14
Qdim, Ordinate	3-20
Qdim, Staggered	3-14
Qleader	3-15
Qleader, Geo. Tol.	3-24
Quadrant, Object Snap	8-5
Quick dimension	3-13

R

Radial dimension	3-11
Radius dimension	3-12
Rectangle	4-44
Rectangular Array	1-2
Redo	1-28
Regenerate Assoc. Dims	3-3
Relative Coordinate	7-3
Rev. Cloud, Convert object	4-48
Revision Cloud, Creating	4-47
Revision Cloud Style	4-46
Rotate	1-24
Running Object Snap	8-8

S

Save a drawing	5-30
Scale	1-25
Scale Copy	1-25
Scale factor	1-25
Scale, Linetype	10-4
Scale Reference	1-25
Scaled drawings	2-3
Scales, Drawing	8-14
Scaling Text	11-12
Select objects	5-31
Selecting Layers	6-2
Settings, Drafting	10-2
Setting Precision	10-7
Setting Pick Box size	10-6
Setting Units	10-7
Single Line Text	11-13
Single Line Text, Editing	11-15
Snap, Polar	7-12
Special Text Characters	11-16
Start a new drawing	5-32
Stretch	1-26
Symmetrical, Tol. Dim.	3-22

T

Table, Create a	5-33
Table, Insert a	5-35
Table, Modify a	5-39
Table Cell, Insert Block into	5-36
Table Cell, Add a Field to a	5-43
Table Cell, Insert Formula	5-37
Average	5-38
Sum	5-37
Tabs, Multiline Text	11-11
Tangent, Object Snap	8-6
Template, Create a	5-13
Template, Open a	5-15
Text Alignment, Dim.	3-46
Text Appearance, Dim.	3-46
Text, Background Mask	8-2
Text, Edit Dim. Text Value	3-30
Text, Geometric Symbols	3-27
Text, In-Place editor	11-6
Text Indents	11-10
Text Justification	11-14
Text, Multiline	11-5
Text Placement, Dim.	3-46
Text, Scaling	11-12
Text, Single Line	11-13
Text, Special Characters	11-16
Text, Editing Single Line	11-15
Text Styles, Creating	11-2
Text Style, Editing a	11-4
Thaw and Freeze (06 & 05)	6-11
Thaw and Freeze (2004)	6-10
Tolerance dimension	3-21
Tolerance, Geometric	3-23
Tracking, Polar	7-13
Trans-spatial Dims	3-49
Trim	1-27
Trim Hatch	4-36
Typing Geometric Symbols	3-27

U

UCS icon, Display	12-2
Undo	1-28

V

Viewports	5-18
Viewport, How to Create	5-18
Viewport, How to Lock a	5-21

W

Wipeout	1-29

X

Y

Z

Zero suppression	3-48
Zoom	1-30